Professionalism and in Law Practice

This book presents practical advice to law students and those entering and now working in the legal profession that will help them to reconcile who they are as a person with the demands and opportunities of a legal career.

The book sets out a clear framework and practice examples for: (i) defining "success", (ii) understanding the role of a professional in relation to clients, colleagues, adversaries and community, (iii) reconciling demands of practice within ethical rules and norms, business considerations and personal values and (iv) building a values-centered, economically viable practice and reputation.

Complete with practical advice and experiences that produce and reinforce a holistic approach, this book provides invaluable support for second- and third-year law students and lawyers in practice to establish elusive work-life balance over the course of a legal career.

Robert Feldman teaches Professional Responsibility and Negotiation Theory and Practice. He was a corporate partner at Weil, Gotshal & Manges specializing in real estate transactions, restructurings and bankruptcy. He was part of the team representing the Chapter 11 Debtor's Estate in both the Enron and Lehman bankruptcies.

Giving Voice to Values

Series Editor: Mary C. Gentile

The *Giving Voice to Values* series is a collection of books on Business Ethics and Corporate Social Responsibility that brings a practical, solutions-oriented, skill-building approach to the salient questions of values-driven leadership.

Giving Voice to Values (GVV: www.GivingVoiceToValues.org)— the curriculum, the pedagogy and the research upon which it is based—was designed to transform the foundational assumptions upon which the teaching of business ethics is based, and importantly, to equip future business leaders to know not only what is right—but how to make it happen.

Engaging the Heart in Business
Alice Alessandri and Alberto Aleo

Shaping the Future of Work
A Handbook for Building a New Social Contract
Thomas A. Kochan and Lee Dyer

Professionalism and Values in Law Practice
Robert Feldman

Giving Voice to Values in the Boardroom
Cynthia E. Clark

For a full list of titles in this series, please visit www.routledge.com/Giving-Voice-to-Values/book-series/GVV

Professionalism and Values in Law Practice

Robert Feldman

Routledge
Taylor & Francis Group

LONDON AND NEW YORK

First published 2021
by Routledge
2 Park Square, Milton Park, Abingdon, Oxon OX14 4RN

and by Routledge
52 Vanderbilt Avenue, New York, NY 10017

*Routledge is an imprint of the Taylor & Francis Group, an informa
business*

© 2021 Robert Feldman

The right of Robert Feldman to be identified as author of this
work has been asserted by him in accordance with sections 77
and 78 of the Copyright, Designs and Patents Act 1988.

British Library Cataloguing-in-Publication Data
A catalogue record for this book is available from the British
Library

Library of Congress Cataloging-in-Publication Data
Names: Feldman, Robert C., 1948- author.
Title: Professionalism and values in law practice / Robert
    Feldman.
Description: Milton Park, Abingdon, Oxon ; New York, NY :
    Routledge, 2021. | Series: Giving voice to values | Includes
    bibliographical references and index.
Identifiers: LCCN 2020031139 (print) | LCCN 2020031140
    (ebook) | ISBN 9780367200107 (hardback) | ISBN
    9780367200428 (paperback) | ISBN 9780429244704 (ebook)
Subjects: LCSH: Practice of law. | Legal ethics.
Classification: LCC K120 .F45 2021 (print) | LCC K120 (ebook) |
    DDC 174/.3—dc23
LC record available at https://lccn.loc.gov/2020031139
LC ebook record available at https://lccn.loc.gov/2020031140

ISBN: 978-0-367-20010-7 (hbk)
ISBN: 978-0-367-20042-8 (pbk)
ISBN: 978-0-429-24470-4 (ebk)

Typeset in Times New Roman
by Apex CoVantage, LLC

Contents

Preamble

A disclaimer: I practiced law for 36 years—at a large regional firm for 14 years and at a very large international firm for 22 years. The observations herein are the product of my personal experience, growth and observations. I believe these observations will provide workable, practical advice to law students and lawyers to help avoid malaise and fulfill the best aspirations of a professional practice of law. Although I have drawn on many sources and writers to form these ideas and strategies, these ideas should be viewed more as a personal memoir presented as a mentor and not a "memoir of the culture" of law practice in a general sense. It is presented with the purpose of providing context and cautions for the law students and lawyers who aspire to implement their highest personal values for their own integrity and to inspire others to do the same.

This book is written with the hope that it will resonate on an individual level with the reader. It is a rock thrown into the water with the ripples rocking others' boats.

Introduction

The purpose of this book is to provide experience-based guidance to individuals contemplating or currently engaged in the practice of law. Lawyers are subjected to many countervailing influences. Balancing those influences is the foundation of achieving a satisfactory professional and personal life.

The structure of the lawyer's career is presented herein with reference to three main categories that must be balanced: (i) the business of law practice leading to sustainable economic and client service goals that provide growth opportunity, (ii) compliance with ethical mandates and professional rules and (iii) consistent achievement of aspirational personal values that serve to create integrity between personal and professional identity. It is the emphasis on creating equality for the third category of "values" that distinguishes this book from similar professional responsibility texts.

A consistent allegiance to a values orientation is, in my opinion, critical to a satisfactory career in law practice. Adherence to values is the path of satisfactorily reconciling the personal and the professional while answering the question "Who am I and who do I want to be as a lawyer?"

In order to serve as a useful guide to enacting the balance of business, ethics and values, the book is organized into four categories. Section 1, *Awareness of Balance Between Business, Ethics and Values in Law Practice*, sets forth the lawyer's obligation to master the law and facts of the matters and cases they undertake, the need to understand and represent the client's needs and wants, the structure of the ethical rules and professional codes of conduct and the need to elevate and express aspirational values that guide our sense of right and responsibility to high ideals.

Section 2, *Analysis and Action in Lawyer/Client Representation and Intra Professional Interactions*, sets forth several critical elements of situations where the lawyer's ability to implement their values are tested. Examples of such tensions and actions that can affect positive outcomes are described.

Section 3, *Core Intangibles–Trust and Honesty*, sets forth observations on the essential emotional components that make law practice satisfactory and sustainable.

Section 4, *Building a Successful Career*, focuses on the tools and leadership skills necessary to achieve meaningful personal growth and contribute to the profession and broader community.

Awareness of balance between business, ethics and values in law practice

Chapter 1

The values-driven professional and *Giving Voice to Values*

All professionals face the challenge of identifying and balancing their individual values with the demands of the profession, clients and economic motivations. Professor Mary Gentile's curriculum and pedagogical approach to values-driven leadership development, *Giving Voice to Values*, based at the Darden School of Business at the University of Virginia, seeks to address these values/business issues in a broader context. This book follows the *Giving Voice to Values* "approach to values-driven action [as] one of alignment, of moving with our highest aspirations and our deepest sense of who we wish to be, rather than a stance of coercion and stern judgement or moving against our inclinations."[1] However, we examine values-driven actions in the specific context of law practice.

A commitment to values-driven action is best implemented by recognizing the competing influences on a professional and implementing strategies to navigate and affect outcomes that create alignment among those competing influences to achieve the professional's aspirations and obligations to themselves, clients, colleagues and family.

A profound observation of *Giving Voice to Values* is to "expect values conflicts so that you approach them calmly and competently."[2] Professor Gentile points out that "often people do not see ethical or moral challenges as natural or integral parts of doing business." "Framing the situation [of our challenges] as extraordinary . . . can sometimes have a disabling effect."[3] A professional should be aware that competing interests are a constant in law practice in large and small ways. The core of the instruction to be gained from this book should be to develop a belief and confidence that values conflicts should be accepted as part of law practice and to build a structure to anticipate and resolve those conflicts in a way that aligns the competing interests with shared values and holds true to personal identity of a values-driven life.

The *Giving Voice to Values* approach to resolving values conflicts seeks to allow an individual to avoid anticipatory obedience to an issue by applying three steps: awareness, analysis and action. Awareness contemplates exposure of the values conflict by recognition of the influences that test and undermine values. Analysis is the process of considering relevant laws, regulations, policies, business considerations and models of ethical responsibility and then assessing the boundaries of such restrictions and guidance for values-driven actions. However, the basic tenet of this book and *Giving Voice to Values* is the recognition that there is an intuitive, emotional response to recognizing what is right and consistent with an individual's values. This is a process of positive deviation—of reviving the intuitive unconscious notion of doing what is right. The action is therefore "post-decision-making." It allows the focus to move from debating what is right, since that is known to the individual, to debating how to do what is right. The process seeks to gain buy-in, to find how to be effective and to implement the right solution that adheres to understood and shared values.

Giving Voice to Values seeks to apply skills and tools through case studies to build "a moral muscle memory" or habit by practicing or rehearsing action plans and possible scripts. The program also seeks to play to an individual's authenticity but doesn't rely solely on righteousness or "moral courage" in order to implement what you know is right. *Giving Voice to Values* is not a prescription but a technique to emphasize strength and modify weakness in taking action to implement a values-oriented solution.

This book is a supplement to the *Giving Voice to Values* program and is focused on the legal profession. I highly recommend further inquiry in *Giving Voice to Values*, as it is not about "can you do the right thing?" but "how to do the right thing."

A law practice will span 30 to 40 years. The credential of a law license opens many different choices and decisions that change and evolve as a professional changes their life situations and career options. However that evolution plays out (including the lawyer's ambition), the definition of success must include incorporation of values-driven choices. Defining such values on a personal level will guide the individual's aspiration to be a values-driven professional. Ambition will be translated to aspiration, in that each challenge will allow your present self to grow into the lawyer you wish to become. This process is an evolution, not a destination, and will be influenced by many factors that will be discussed herein. Although this journey of aspiration requires personal strength and commitment, it is not a singular enterprise but relies on colleagues and mentors as well as individual efforts. We will discuss the guideposts and resources to promote the goal of achieving success as a values-driven professional as you will define and refine it.

A word of caution is appropriate before embarking on the examination of values-driven professionalism. As a person and as a lawyer, a professional must listen and understand their clients, colleagues and family, and incorporate the values and needs of these individuals into their own actions. They must create a balance between honoring the values of clients, colleagues and family and honoring their own values. A lawyer's primary goal is to understand the client's objectives and values, and in the context of the issues presented, to negotiate an alignment of values. Values should not be used arrogantly or as a shield to avoid a professional's legitimate duties. The challenge of a successful professional career and a satisfying life lies in creating this balance.

Balance does not require compromise of values but rather assertion of your values in a manner that fully integrates values into actions that implement ethical and business decisions. By acknowledging the importance of values, the individual asserts their integrity and their desire to be a whole person, balanced in serving their and the client's need for business, ethical *and* values-oriented goals when reaching desired outcomes.

Notes

1 Mary C. Gentile, *Giving Voice to Values: How to Speak Your Mind When You Know What's Right* (New Haven, CT: Yale University Press, 2010), 28. www.GivingVoiceToValues.org/
2 Mary C. Gentile, *Giving Voice to Values: How to Speak Your Mind When You Know What's Right* (New Haven, CT: Yale University Press, 2010), 72. www.GivingVoiceToValues.org/
3 Mary C. Gentile, *Giving Voice to Values: How to Speak Your Mind When You Know What's Right* (New Haven, CT: Yale University Press, 2010), 73. www.GivingVoiceToValues.org/

Chapter 2

Character and values

Sound decisions flow from values. As described in *Giving Voice to Values*, the term "values" is an "overdetermined word." For our purposes, "values" has a moral dimension based upon an intuitive, internally generated sense of right and wrong. Widely shared values are honesty, respect, responsibility, fairness and compassion. "Values" should be distinguished from two other guiding principles: "ethics" and "business." Together these three factors form the basis for decision making for a lawyer.

Ethics in the professional context refers to various external frameworks that can help discipline our thinking about various ethical choices and dilemmas. In particular, discussions of ethical reasoning often focus on scenarios where various models will lead us to different, conflicting decisions about what 'is right,' and what follows may be a discussion of how we might try to resolve those tensions.[1] Lawyers subscribe to a system of rules or standards with which one is expected to comply; failure to do so carries various penalties. "Thus ethics is often seen as rule-based and externally imposed, something that exists outside the individual."[2]

The career and economic considerations described herein as "business" are meant to recognize that a lawyer in society is part of the economic machinery of commerce. Legal careers take many forms, but lawyers in practice are supporting themselves as well as their families, their colleagues and their communities in many ways. The building of reputation, client base and fiscal security are powerful motivations. Most decisions presented to a lawyer in practice contain conflicting objectives and undiscovered options requiring predictions of future legal and economic outcomes at varied levels of uncertainty and risk.

The role of the lawyer as a counselor will be discussed in Chapter 8. The counseling function of law practice involves influencing the client's decision matrix. However, the injection of the individual's values into the decision matrix of values, ethics and business is the recognition that "character" is an essential component of a satisfying and values-driven law practice.

The definition of "values" as used in this book relies heavily on the description of values in Chapter 2 of *Giving Voice to Values*.[3] Simply stated, these are not exclusively arbitrary individual values but rather a core set of values that tend to be universally shared across cultures and time (although, of course, not for every individual). For example, Rushworth Kidder identifies five widely shared values: honesty, respect, responsibility, fairness and compassion.[4]

"Character" is the quality of recognizing and setting the frame of an individual's values on the decision process. "Integrity" is the state of being whole and undivided. Character and integrity combine to create your moral compass, which helps direct the action that is right for you to implement your values.

The popular idea of "thought experiments" proposes that an individual assess a situation ripe for a decision by examining the consequences of various actions and omission of actions. For example, imagine you are in a meeting where actions are proposed that may not be illegal but clearly violate your sense of right and wrong when it comes to the values that are critical to your sense of well-being and integrity. Do you speak up in the meeting? Do you wait and address the offending parties or others after the meeting to correct the situation to your satisfaction? Do you let this circumstance pass but plan strategies to correct at a future date? Each of these alternatives are a test of character and each may have compelling reasons to be right for you. However, the key is recognition and awareness of the need to integrate your values analysis into the decision matrix for you and others.

The expectation that this values component will be a constant for your conduct, and your strategies to address it, will lead to strength of character and pursuit of individual integrity.

There will always be a balance of values, ethics and business in law practice. Finding that balance and providing appropriate weights for each factor while preserving integrity and considering the needs of clients, colleagues, family and community should be the aspiration of the values-driven professional.

As described in the Introduction, the *Giving Voice to Values* approach first requires determining the values relevant to the issue presented and then focusing on action. In other words, the conversation with colleagues and clients moves the discussion from what is right to "how do we accomplish what is right?"

The *Giving Voice to Values* technique is first to anticipate the impending conflict, thus extending the time to prepare, and next, to prescript responses and allow the values-oriented action to default to an informed voice. Then the third step, action, is critical to this process.

Recognizing that values conflicts are a constant allows for awareness and anticipation. Participants in a values-conflict decision can be divided into "idealists, pragmatists and opportunists."[5] The following definitions may be useful: "idealists" attempt to act on their moral ideals no matter what; "pragmatists" seek a balance between their material welfare and their moral ideals and "opportunists" are driven exclusively by their own material welfare. Pragmatists usually try to do the right thing. Opportunists seek to find the solution that suits their (lesser) values component. Examples of opportunism would be: (i) price-gouging shortages of necessary supplies of food, water or energy, (ii) requiring non-disclosure of a dangerous condition or product as a component of a legal settlement with an individual complainant, thus putting others at risk or (iii) employment practices that discriminate or exploit. These characterizations (idealist, pragmatist, opportunist) are general tendencies, self-identified, rather than a description of what one does all the time. They are not "fixed" but rather inclinations toward certain motivations/priorities. By applying the techniques and awareness described herein and in *Giving Voice to Values*, both the pragmatists and the idealists can change the calculus that the opportunist uses to determine what is in their self-interest in a given circumstance.

The pragmatist should try to avoid anticipatory obedience to the proposed or existing practice and create awareness of the "right" values-oriented choice. The pragmatist affects the ability of the opportunist to control the decision because by implementing the pragmatist's informed voice and prescripted rationale, the pragmatist does not allow the opportunist the ability to operate with impunity. This approach will elevate the discussion from "should we continue or implement the values challenged practice?" to "how do we accomplish a result that incorporates commonly held values?"

The professional has different and additional constraints. The solution not only needs to satisfy the values judgment of the professional but also must satisfy the client's economic and emotional needs and the rules of legal ethics.

The application of these factors will be determined by the character and preparation of the professional. The goal should be to practice moral competence and adapt solutions and actions to your individual strengths while building the moral muscle to improve the recognition and persuasion skills necessary to implement the values component consistent with ethical and business considerations.

Notes

1 Mary C. Gentile, *Giving Voice to Values: How to Speak Your Mind When You Know What's Right* (New Haven, CT: Yale University Press, 2010), 25. www.GivingVoiceToValues.org/
2 Mary C. Gentile, *Giving Voice to Values: How to Speak Your Mind When You Know What's Right* (New Haven, CT: Yale University Press, 2010), 25. www.GivingVoiceToValues.org/
3 Mary C. Gentile, *Giving Voice to Values: How to Speak Your Mind When You Know What's Right* (New Haven, CT: Yale University Press, 2010), 24–46. www.GivingVoiceToValues.org/
4 Mary C. Gentile, *Giving Voice to Values: How to Speak Your Mind When You Know What's Right* (New Haven, CT: Yale University Press, 2010), 30. www.GivingVoiceToValues.org/
5 Mary C. Gentile, *Giving Voice to Values: How to Speak Your Mind When You Know What's Right* (New Haven, CT: Yale University Press, 2010), 109. www.GivingVoiceToValues.org/

Chapter 3

Professionalism beyond legal competence

There is an important difference between an "expert" and a "professional." An expert utilizes their knowledge of a subject to see the application of facts and circumstances to various alternative solutions or outcomes. The expert has free rein to use their knowledge of the discipline they have studied without the demands of others. On the other hand, a professional must first be an expert on the matters they are dealing with but must also take into consideration the needs and desires of a client. The introduction of the client distinguishes an expert from a professional. Historically, professional societies have been charged with not only specialized knowledge but also a moral commitment to the public good.[1]

As we will discuss in further detail, there are other important considerations that a lawyer should take into account in order to achieve a fulfilling life in law practice and meet the expectations of the profession. In addition to (i) considering their expertise in the applicable law and legal principles, (ii) carefully ascertaining the facts and (iii) fully communicating with the client regarding the client's needs, the next step in the lawyer's decision making should be to set their perspective on the situation at hand in a way that aligns with the values and the professional responsibility of the lawyer. The frame established by perceiving the situation and the available options is not just a transparent way of absorbing the law, facts and client's expressed desires but rather is a way of applying the intuitive and carefully considered sense of right and wrong that has been developed by the lawyer over their lifetime. The triggering of the judgments and responses in the lawyer's mind, gained over the lawyer's lifetime and integrated with universally shared values, will influence the lawyer's perspective and actions. The application of those unconscious judgments that have come from the lawyer's life experiences in their family, school, community and work life is "character." Each experience builds, shapes and tests the lawyer's character, so that they may act when they know what is right.

The aspiration to insert the lawyer's character into every professional encounter is the foundation of a values-driven law practice and alignment of the lawyer's life goals with their legal ambitions.

Four factors are critical to the successful application of this approach. First, the lawyer must be diligent and competent in thoroughly mastering the law and facts as presented to them by the client. Also, the lawyer must patiently listen to all of the client's needs and assess the demands of counterparties to the transaction or litigation.

Second, the lawyer needs to set the frame of representation and access their intuitive character to predict a positive outcome that will guide the lawyer's actions consistent with the lawyer's values. The predicted positive outcome will reflect an aspirational result that will align who the lawyer wants to be as a person with how they will practice law.[2]

Third, the lawyer should script the merger of their legal, factual and situational analysis with their aspirational character and communicate that integration with the client, their colleagues and counterparties as the situation evolves and changes from time to time.

Fourth, it is critical that the lawyer be flexible and communicative as circumstances dictate but stay true to their aspirational character and seek to find ways to align the demands of the client and others. This awareness has been referred to as the "learning to see" model.[3] These changes and challenges as the representation progresses are opportunities to build and strengthen character. Good character emerges out of the mysterious interplay of many little good influences. However, in all likelihood, all will not go smoothly. Willful clients, colleagues and counterparties as well as family and other allies will all have agendas that will test the values-driven lawyer's ability to implement the script or frame set at the beginning.

Recognizing and adjusting for these counter-influences and continuing to strive for an acceptable level of balance is very important. Preparing and anticipating this resistance allows the lawyer to approach them calmly and confidently. Overreaction can limit your choices unnecessarily.[4] It is in this fourth element, values conflicts among participants, where the lawyer must integrate their personal values with their professional obligations. If you recognize that values issues can have equal status with legal or business issues, it will seem more natural to advocate for inclusion of values strategies alongside business and legal issues. In other words, don't compartmentalize values from other aspects of professional advice. Lawyers are taught to explain or argue their positions on contrary legal points, risks of business strategies or factual disputes; however, when it comes to value judgments, they sometimes defer in speaking up, feeling that their personal values should not be part of the discussion. Certainly economy of thought is appropriate in advocating on any aspect of a representation; legal, factual or business considerations should be carefully considered. However, failure to include integration of core norms and universal values that have been consistent across time and professions[5] is unprofessional, and expressing those views when you know what is right is critical to achieving a satisfactory professional life and serving the best interest of your client, family, community and yourself.

An important caveat here, as you likely already know, is that you may face tough choices and strong resistance by expressing values that clients or colleagues feel may jeopardize business, personal or legal outcomes. The application of the *Giving Voice to Values* approach of anticipation through awareness of the impending conflict, prescripting responses based upon many of the ideas contained in this book and acting with character will aid in successfully addressing resistance.

In complex decision making, weighing legal or factual issues may seem easier to discuss as core elements to creating outcomes and alternatives. Lawyers are taught legal analysis, application of case law precedent and legal theory. Factual analysis may have similar analytical or scientific bases which are commonly debated or processed to arrive at conclusions. Values are more emotional and intuitive, and may be subject to rejection as not expedient or necessary to the analysis. The values outlined in the *Giving Voice to Values* approach are a consistent foundation that has existed across time and cultures, observed by philosophers, psychologists and other researchers, and they have allowed societies to function.[6] It is therefore important, recognizing that values conflicts constantly present themselves, to give the values analysis equal dignity in the discussion and, pursuant to the *Giving Voice to Values* technique, an authoritative, certain voice in framing that awareness and analysis as an essential component of the course of action to be recommended and/or undertaken.

You also likely won't win every battle and will have to make tough decisions on when and how to speak up. Subsequent chapters will discuss strategies for mitigating these decisions while remaining true to your values.

The final, and critical, point is that most battles, even when they end in "failure," are valuable in building and strengthening judgment.

The previous discussion can be illustrated in the following example.

A husband and wife have filed for divorce. They have two school-age children. The husband has a significant income; the wife is the primary parent in raising the children and does not work outside the home. The husband expresses to you, his attorney, that he has a new girlfriend with whom he wants to travel and spend time on the weekends, and he is totally indifferent to any visitation or time with his children. He tells you he understands that it is a common tactic for husbands in his situation to file for child custody and/or disruptive visitation in order to prolong the time and legal costs of divorce, which the husband can more fully sustain, and scare the wife with threatened loss of custody in order to cause the wife to settle for lower child support and lesser property settlement. You have seen this tactic enacted in other divorces, achieving the outcome the husband suggests, but you know that the children suffer from lesser life options and emotional issues from the husband's "abandonment" of the children and their lack of relationship with their father.

Subsequent chapters will provide tools that will allow the husband's lawyer to represent the best interest of the husband but maintain the lawyer's sense of the right course consistent with the lawyer's values. If the husband seeks to manipulate the custody/child support dynamic of litigation, the lawyer—in implementing values of honesty, fairness, responsibility and compassion—can seek to provide counseling to the husband that finds a values-driven result.

The action to be taken should be as follows. First, at the initial meeting, the lawyer should fully understand the husband's motivations, life situation, financial capacity and goals, including external pressures and considerations such as the influence of the girlfriend, the parenting skills of the wife and her support system. The lawyer should explain that balancing advocacy for the husband with the lawyer's ethical duties, values and experience in the tactical use of litigation (especially in the family law context), requires consideration of the best interests of the children and family, which is a core value of an attorney in family law practice.

In determining the objectives of representation, the lawyer should establish long-term and short-term goals with the client. The lawyer should also explain the duties of loyalty to the client as influenced by the duty of honesty/candor to the court. Misrepresenting the husband's true intention will have consequences beyond the possible potential short-term monetary savings. The lawyer may cite their past experiences regarding the long-term consequences and emotional damage to both the husband and the children that can occur in this dynamic. Importantly, the lawyer should be sure that the counseling relationship allows for collaboration where the lawyer's values and experience can be understood, adopted and utilized by the client. In these situations, the ethical rules (as described later) provide the attorney the opportunity to marshal family counselors or psychologists to help the husband assess the consequences of his actions. The lawyer should help mitigate the short-term, often emotional, dynamic of money and disappointment of divorce with the longer-term perspective relative to the interests of the family and responsibility and compassion for all involved.

The character necessary to resist the client (and those that may influence the client, e.g. the girlfriend) is the test of the lawyer who wants to serve an aspirational values practice. The husband may see the lawyer as a tool to accomplish a small result. It is the values-driven lawyer's challenge and duty to serve the greater interest of the client by providing perspective on the long-term relationship and damage to the client and the children. The lawyer's commitment to honest advice will be the path to building trust and gaining collaboration on shared values.

Notes

1 Deborah L. Rhode, David Luban, Scott L. Cummings, and Nora F. Engstrom, *Legal Profession*, 7th ed. (St. Paul, MN: West Academic, 2016), 36.

2 American Bar Association Center for Professional Responsibility, *Model Rules of Professional Conduct* (Chicago, IL: American Bar Association, 2019), Rule 2.1.
3 David Brooks, *The Social Animal: The Hidden Sources of Love, Character, and Achievement* (New York: Random House, 2012), 128.
4 Mary C. Gentile, *Giving Voice to Values: How to Speak Your Mind When You Know What's Right* (New Haven, CT: Yale University Press, 2010), 72. www.GivingVoiceToValues.org/
5 Mary C. Gentile, *Giving Voice to Values: How to Speak Your Mind When You Know What's Right* (New Haven, CT: Yale University Press, 2010), 24. www.GivingVoiceToValues.org/
6 Mary C. Gentile, *Giving Voice to Values: How to Speak Your Mind When You Know What's Right* (New Haven, CT: Yale University Press, 2010), 24. www.GivingVoiceToValues.org/

Chapter 4

Legal ethics and duties

As described in the prior chapter, there are levels of duties a professional should consider in their practice. "Legal Ethics" is typically defined as the externally codified standards of practice and attorney conduct that are formulated by state regulation, bar association codes and opinions and court decisions. Law firms, companies or government offices also impose policies, rules or requirements regarding acceptable practices and conduct. The consequences of violations of these rules may result in sanctions, including prohibition or suspension of the right to practice law.

Subsequent to the corporate scandals and fraudulent corporate conduct leading to the collapse of WorldCom, Enron and Global Crossing among others, law schools and professional organizations have mandated compulsory training in Professional Responsibility. As part of the bar admissions process, bar examiners include knowledge of the legal ethics and a character review as part of the bar examination and licensing process.

It is important to note that the various codes of Legal Ethics contain both (i) restrictions on unacceptable conduct and (ii) positive duties to clients, courts and the legal profession and community. The explanation and comprehensive analysis of the Code of Professional Responsibility[1] as pronounced in the various sources of the rules are not the subject of this book, as there are excellent sources available for such study. Also the approach of a values-driven law practice is directed to aspirations that serve a purpose that is less about sanctions than about the desire to serve in an exemplary manner to inspire basic values of honesty, respect, responsibility, fairness and compassion.

Professor Carolyn Plump in *Giving Voice to Values in the Legal Profession* observes that the "rules-based emphasis" on legal ethics codes requires not only "knowing the rules but understanding their purpose."[2] Setting a positive frame for understanding the role of ethics in conjunction with values and business requires embracing the lawyer's duties to serve their clients, the court and the profession. The comments and ethics opinions that interpret the Code of Professional Responsibility often seem like escape hatches to avoid imposition of the rules in a manner that might restrict only sharp practices on marginally acceptable conduct. Effective advocacy and zealous representation can be reconciled with ethical responsibilities and duties if values are given equal weight in balancing considerations of ethics, values and business.

Legal ethics should not be seen as a minimum standard to be compromised in pursuit of a lawyer's personal goals or the demands of a client or others that seek to override values to accomplish goals that violate defined values. The key is a consistent expectation that the balance can be maintained and that clients, colleagues, judges and other constituencies can find a common basis for resolving issues with a balanced, clearly communicated blend of these three elements: ethics, values and business.

The International Bar Association (IBA) was founded in 1947 to unite legal practitioners and law societies on a global basis extending to 160 countries. In 2011, the IBA adopted the International Principles on Conduct for the Legal Profession. These legal principles serve as a simple guide summarizing the requirements all ethical attorneys should pursue:

1 Independent, unbiased professional judgment.
2 Honesty, integrity and fairness.
3 Avoidance of conflicts of interest.
4 Confidentiality/professional secrecy.
5 Clients' interest should be paramount.
6 Lawyers should finish what they start.
7 Clients may choose their lawyer.
8 Accountable for held property.
9 Work in a competent and timely manner.
10 Fees must be reasonable and billable work must be necessary.

Some of these principles will be specifically discussed in later chapters. However, the Duty of Competence, which is Rule 1.1 of the Model Rules of Professional Conduct (hereinafter referred to as the "Model Rules"),[3] requires special mention here. The rule states that "competent representation requires legal knowledge, skill, thoroughness and preparation reasonably necessary for the representation." There is no higher ethical duty than a full understanding of the facts and the law relevant to any legal engagement. Without understanding of the facts and law, the lawyer cannot adequately advise or pursue strategic alternatives and communicate choices to their client. The client is entitled to be represented by an attorney who is prepared, engaged, timely and open to understanding the unique circumstances of the situation and the client's needs and objectives. This is the baseline from which all other actions must proceed.

Notes

1 American Bar Association Center for Professional Responsibility, *Model Rules of Professional Conduct* (Chicago, IL: American Bar Association, 2019).
2 Carolyn Plump, *Giving Voice to Values in the Legal Profession: Effective Advocacy With Integrity* (London: Routledge, 2018), 28.
3 American Bar Association Center for Professional Responsibility, *Model Rules of Professional Conduct* (Chicago, IL: American Bar Association, 2019).

Chapter 5

Reconciling character and values with expectations of clients and employers

Law practice is both a profession and a business. As a professional, the lawyer uses their knowledge and skills pursuant to the standards of ethics and applies their values to affect outcomes for their clients. As a business, the lawyer's success is measured by their effectiveness in creating desired outcomes for their clients and the legal organizations the lawyer serves. The lawyer's character is constantly challenged to reconcile the proper balance between the lawyer's role in the profession and the business.

Ethics are external rules and standards of conduct set by the legal establishment (the bar, the courts and government agencies as well as accepted standards of practice). Values are the internally generated standards of conduct of individuals, like-minded associates and the community. Legal professionals, for themselves and their clients, are at the intersection of conflict and deal-making that tests the ability of people to compromise and resolve differing social and economic objectives, often ferociously and with great passion. Character is the ability of the legal professional to stand up for the balance of ethical, legal and business conflict and compliance.

The legal professional can expect that clients and employers, eager to obtain desirable personal and/or economic outcomes, will often seek to influence the conduct of the lawyer to favor business or professional outcomes over values and/or ethical considerations.

When a legal professional is called upon to engage in unethical conduct, clients and employers will grudgingly accept that the external rules that could lead to sanctions affecting both the lawyer and the client or employer provide sufficient reason for the lawyer to refuse to cross ethical "red lines." On the other hand, when a lawyer asserts a values conflict (an exercise of character), that assertion is often viewed as a personal decision and may even be deemed an expression that is unacceptable to the notion of loyalty to the idea of "zealous representation" or "zealous advocacy." The recognition of the values choice and elevating values to an equal position in the balance of values, ethics and business is the first step in creating the mindset or frame for exercising character. Unless a lawyer can clearly articulate the value that is being challenged, they cannot begin to marshal the resources to achieve values-oriented results.

In the prior example of the divorcing husband who tactically wants to threaten a custody battle to reduce the child support award, the lawyer must explain and assert the purpose of the domestic relations laws to give primacy to the best interests of the children through the father's personal and financial support. The lawyer must also assert their understanding of the father's wishes moderated by the lawyer's sense of responsibility, compassion and honesty in the court proceedings and negotiations. The assertion of these values makes the lawyer more than a tool in the husband's monetary goals. The lawyer can thus serve the best interests of the client and express the values of the lawyer and client as well as provide the context of the theory of public policy (i.e. the best interests of the children) embodied in the applicable law.

We will discuss honesty in more detail in Chapter 11, but an example here may better illustrate the issue. Model Rule 4.1, Truthfulness in Statements to Others,[1] prohibits a lawyer from making a false statement of a material fact to a third person. However, Comment 2 to Rule 4.1 exempts "generally accepted conventions in negotiation. . . [including] estimates of price or value placed on the subject transaction." Because all parties recognize the ethical duties of a lawyer, parties will often (reasonably) expect the lawyer to act and speak honestly and not solely be the instrument of the client's economic motivation. The value of this position of apparent relative fairness and honesty is useful to all parties seeking a fair process and outcome. However, it is common in business transactions or litigation settlements for the client to ask the attorney to overstate (lie) about other offers or acceptable settlement amounts. "Tell the buyer that I have an offer 120% higher than his offer," even where no such offer exists. Also "tell the other side I will not settle for any less than $1 million" when the authorized settlement amount is actually $750,000. The ethics rules and business conventions minimize the effect of this dishonesty if the lawyer succumbs to the client's instructions.

On the one hand, the lawyer can seek to persuade the client of the bad business outcomes (withdrawal of prior offers, legal recourse for fraud) if the other side discovers that this was, in fact, deceit. Or the lawyer can appeal to the reputational harm the lawyer or the client may suffer from the discovery of the deceit. But the client may persist.

The dynamics of employment for in-house counsel and junior lawyers in a law firm or legal organization may provide the lawyer with limited recourse in these circumstances.

Therefore, how should a lawyer seek to mitigate the situation?

First, recognize the value at issue: personal honesty. Contrary to popular perception, I believe that a lawyer may go through their whole career without lying, but it requires character and resolve.

Second, as part of the preparation process (i.e. very early in the representation), identify this value as a part of your representation and confirm that the client and colleagues share that value. As earlier described, expect that you will encounter ethical and values conflicts from some of the constituents to the dispute or transaction so that you are able to anticipate how to respond in an honest fashion that is consistent with your client and employer's expectations but is additive to their business objectives. Don't succumb to the lowest common denominator of conduct displayed by other parties to the litigation or transaction.

Third, assess both your position based upon ethics and values and the demand/request being made by your client and/or employer to ascertain how you can best serve the objective of the representation. This analysis allows you to best represent your client's objectives and implement your values without retreating to a moral stringency that overstates your concern. In other words, your duty to the client's goals don't change but rather your approach to how or if you implement their requests should be tailored to include an expression of your values. As described in later chapters, Rule 2.1 of the Model Rules[2] allows an attorney, as an advisor, to express guidance on a range of factors including "moral, economic, social and political factors that may be relevant to the client's situation." However, the expression of the lawyer's values in those circumstances must take into account the client's and employer's values in a manner that tries to find an acceptable, values-driven path forward.

For example, an employer of undocumented workers could consult a lawyer regarding how to plan for the economic and legal jeopardy that they may face in anticipation of mass enforcement actions by Immigration and Customs Enforcement officers who have raided similar businesses in the area. If both the client and the lawyer share a desire to address the human dislocations that will impact the undocumented workers, the legal liability and economic displacement to the employer and the other employees and customers of the business, the exploration of common values and goals will be a starting point for devising a legal strategy to address the legal, economic, moral and political factors at issue. The lawyer with experience in how other employers have addressed these matters and the solutions that might be available through the Immigration Agency's best practices and policies would be providing legal advice with a values component consistent with the lawyer and client's shared values and the best interest of the affected constituents, including shareholders, customers and workers (documented and undocumented).

Anticipation is often the product of experience. As a young lawyer, you may expect ethical and values challenges but not recognize them until they are on top of you. Access the knowledge and guidance of more experienced lawyers who may have encountered some of the actors or experienced some of the circumstances you are facing.

It may be useful to elaborate on the importance of preparation to avoid surprise and reaction to values conflicts. Many cases or transactions will fit predictable patterns with milestones where decisions need to be made. A lawsuit will progress through discovery, summary judgment, pretrial motions and trial with particular inflection points based upon the facts, law, venue and parties. An experienced practitioner will be able to predict pressure points, settlement opportunities, schedule and projected costs for each stage. The litigation plan, schedule and budget are all discussed and evaluated at various stages of the case. A seasoned practitioner can also predict and discuss with the client and experts the timing of revelations and decisions that must be made as circumstances unfold in the litigation and judicial decisions or as discovery affects the adjustments to the plan. Besides litigation tactics, as part of setting the objectives of the representation, the lawyer should also reserve a portion of the discussions to check in on the congruence of values regarding honesty, economic versus social goals and fairness/compromise versus seeking maximum capture of any potential recovery. Although these topics may have an abstract quality prior to the presentation of the actual decision in the case, the prior discussion will serve as a touchstone for the later, more precise, discussion.

Experience is the best predictor of where these conflicts may come up, and anticipation is helpful but never a complete solution for what is sometimes a hard discussion of the values element that may affect economic, tactical and ethical solutions. If these conflicts occur, especially early in your career, allow yourself space to make mistakes, admit them, get help to resolve them and move forward.

Notes

1 American Bar Association Center for Professional Responsibility, *Model Rules of Professional Conduct* (Chicago, IL: American Bar Association, 2019).
2 American Bar Association Center for Professional Responsibility, *Model Rules of Professional Conduct* (Chicago, IL: American Bar Association, 2019).

Section 2

Analysis and action in lawyer/client representation and intra professional interactions

The following four chapters represent the ethical rules from the Model Rules contained in the Code of Professional Responsibility[1], which provide the structure for formulating the action plan that will build alignment of values between lawyers and clients. These rules also create the alignment among colleagues and adverse professionals to cooperate in finding values-oriented action paths, which can move the resolution of cases and transactions forward.

Chapter 6

Who is the client?

Earlier discussions identified that a key difference between an expert and a professional was that a professional uses their expertise (consistent with the ethical rules and values) to serve the goals of a client. However, the person communicating with the professional regarding the tasks and services to be rendered to "the client" is often not the client but a representative of the client. The representative also often has needs and agendas that may not be fully consistent with the objectives of the "client."

As simple as it may seem, ascertaining who is the client (and who isn't) is often a threshold issue. Lawyers are required to define the terms of their engagement in a written document—an "engagement letter"—at the start of the representation. It is important to clarify who the client is and how instructions and communications should proceed in the engagement letter.

A young lawyer often communicates with their supervising attorney regarding client needs. A business entity communicates with the lawyer through officers and agents. Family members may express the intentions of "the family" to the lawyer, purporting to represent the wishes of all members of the family.

Chapters 6, 7, 8 and 9 will unpack the practical application of client communications and representation as well as the ethical rules that apply, including setting objectives of representation, modes of client counseling and conflicts of interest.

Although a lawyer may receive instructions from or have contact with multiple parties who purport to explain the "client's" objectives, the lawyer must fully inquire about the role and potential conflicts or self-interest of each party to the litigation or transaction.

Over the course of a lawsuit or a transaction the interests of the parties may shift. It is very useful to consult with allies regarding potential paths that each party may take based upon those allies' past experiences in similar litigation or transactions. Identifying potential changes in positions can help avoid difficult ethical and values decisions when positions or time demands shift.

For example, officers of a business entity are often compensated on the successful outcome of a transaction regardless of the long-term benefits to the company. The lawyer does not need to make predictions as to long-term success or failure of a transaction to recognize that the person instructing them regarding legal actions to be undertaken may not have aligned objectives with the shareholders or investors who receive other forms of compensation for their investment.

If the business entity is the client, as opposed to one or more of its constituents, the lawyer needs to remain clear with the business entity's representative about the lawyer's loyalty to the goals of the business entity.

Sometimes all parties, officers, investors and the company have separate counsel to advocate and compromise their respective interests and the legal steps to be taken to document and implement those interests. However, more often than not those separate representations fade away after entity formation, and the independent "lawyer for the company" is left to look after all potentially competing interests. There must be clarity about where the lawyer's responsibilities reside (usually with the entity) when suggesting courses of action that may affect executive compensation to the corporate officer (who has hired the lawyer) or the corporate in-house counsel (who owes their employment and compensation to such executive). Otherwise, it becomes very difficult for the outside lawyer to take a stand regarding the primary duty of the lawyer to the entity—not to the executive who is the representative of the company.

The tension between these conflicting positions can be mitigated by clearly outlining the boundaries of the "client" and the lines of communication and authority to instruct the lawyer. However, consent alone—although usually sufficient to accept instructions—does not absolve the values-oriented attorney from questioning authority and motives. Legal advice should not be a path around improper or illegal actions. Exploring objectives with a client can aid in devising an effective legal, ethical and values-oriented plan. Of course, tact and experience will help avoid confrontations, but the steps for client interactions outlined in this section may help establish clear communications to achieve the balance between values, ethics and business.

As a consistent practice, every engagement should be preceded with an engagement letter that contains the following items:

1   Define whom you are representing with due regard to corporate entity names (including specific subsidiaries, if applicable) and the role of officers, directors and affiliate parties by title and/or name. This may need to be amended from time to time.
2   Define fee arrangements and billing procedures.
3   State which lawyers, in addition to yourself, will be working on the matter.
4   Define the scope of the representation and, usually, disavow that this is general representation of all legal matters.
5   To the extent feasible, clear conflicts of interest that may occur or are known as of the date of representation for unrelated matters.
6   Identify the process to schedule and define objectives of representation and the obligation of both parties to communicate and revisit such schedule and objectives.
7   State the terms of any retainer.

If more than one lawyer is involved in the representation, there should be an internal process to coordinate advice and consultation to "speak with one voice" for the benefit of the client. The goals of team members (often for business development purposes) can impact the advice given.

Typically, law firms have a form engagement letter with prescribed language that members of the firm are asked to include with minor variance. As with all forms, this can be problematic if care is not taken to be sure the form accurately reflects the circumstances of each particular retention. General language left in a form can be contradictory on many levels and should be reconciled with negotiated terms.

A common situation in commencing representation involves the desire to rush through the mandated procedures for establishing the client relationship with a new engagement letter tailored to the particulars of the client and the new representation in order to begin to solve the issues presented by the new representation. The procedures are often viewed as a necessary but inconvenient gateway by both the lawyer undertaking the matter and the client seeking immediate answers and help. Thus, the path of least resistance is to try to force the standard letter through the approval process for new clients or new representations at the firm or try to persuade the client that the mandated language is "boilerplate" and should be executed promptly so work on the new matter can proceed. Like all matters and actions an attorney will undertake, the establishment of the terms of representation needs to be precise and well considered, just as the actual legal work on the new matter would be. This is a good opportunity for the client and lawyer to establish their relationship and demonstrate to the client, and the firm, the desire for compatibility between the client's and the lawyer's shared goals and values, including attention being paid to detail in all aspects of the representation.

Therefore, the lawyer must carefully review and establish the client's concurrence with the terms of the engagement letter and make amendments or additions to the approved form as necessary to carefully express the relevant terms of each particular representation.

It is poor practice to ever commence representation until all identified conflicts have been cleared, pursuant to established conflict clearance procedures and with the clearance communicated and approved by the prospective client.

The lawyer needs to establish the value of candid communication between the primary client contact and the team of lawyers. The agreement of this role of candor should be identified at the start of the representation to help establish the attorney's role as a professional.

Clearly, if you are representing an individual, the lines of communication are simplest. However, most complex cases or transactions have multiple actors both on behalf of the client and as advisors to the client. The lawyer's principal role of defining the facts and applicable laws and advising the client on predictable outcomes can be distorted or influenced by the agenda of individuals or advisors. The professional may have to call out potential conflicts or opportunists who seek to influence the definition of the facts and/or law in a way that is advantageous to the opportunist's interest. The confrontations that may occur when the unbiased professional identifies these issues candidly (and hopefully with poise and tact) are a challenge to the value of honestly expressing what the professional observes as the correct outcome consistent with the objectives of "the client." The professional need not resolve the conflicting objectives, but they should present the conflict so that the parties have sunlight on the identified issues to be resolved.

A common scenario involves the situation where the compensation of an outside advisor (e.g. a broker) is based upon the completion of the transaction (e.g. a sale or financing) with no continuing involvement in the successful long-term outcome of the transaction. The same can be true of compensation internal to the client by executives who are paid (through bonuses or other arrangements) based upon the completion of a transaction rather than the long-term success of the company and its investors. The former situation is easier to tactfully identify than the potential for conflicting objectives of internal parties, some of whom may be directing the attorney's actions. In these circumstances, an experienced lawyer will seek to identify all constituents within the company (managers, executives, general counsel and investors) and establish early and consistent dialogue to allow each constituency to assert their views on the balance between short- and long-term company objectives relative to the subject transaction.

It is important that the client understands, from the very outset of the representation, that the attorney will adhere to this duty of candor to the client's objectives. It is also valuable to identify the ultimate arbiter (e.g. the company president, the general counsel or a committee of the Board of Directors) of any such disputes.

The issues discussed in this chapter can be illustrated in the following example. A large media company that produces family-oriented content has several divisions: (i) theme parks, (ii) movies and television content, (iii) news and (iv) consumer products. Each division has a legal staff embedded in the division's offices that works closely with the division's employees. Each division sets the compensation of their employees based upon meeting revenue production goals. The embedded legal staff interact daily with the division's executives and staff on issues of product safety, news content, themes and portrayal of characters in movies and television productions. The lawyer's advice reflects appropriate interpretation of intellectual property, consumer safety, objective news coverage, racial, sexual or social issues and other legal advice. The boundaries of that advice will be pressed by a business person who sees the outcome as affecting their career or their salary, bonus or advancement based upon production of revenue from the project proposed by the executive. The lawyer's client is the company, not the executive. If the lawyer's "boss" (i.e. the person who determines compensation and advancement of the lawyer) is the executive, the lawyer will find it more difficult to provide advice that prioritizes and upholds the company's reputation as a family-friendly company. However, if the lawyer's boss is the general counsel, and the company policy invests final decisions on the inclusion of the company's values in all particular legal advice, then the lawyer may enlist the general counsel as the arbiter and ally in their advice to protect the true client, the company.

Assume the executive in charge of consumer products at this company is under pressure to produce merchandise depicting characters from a movie that is very popular with young girls (ages 4–12). A designer produces a doll that is more sexualized than the character, thus in conflict with company policies regarding such depictions. However, the pressure of timely distribution and meeting sales goals pushes the executive to authorize the manufacture of the doll. The executive is ready to green-light the release of the doll when a concerned employee alerts the in-house lawyer that the doll violates the company's family-orientation guidelines. The lawyer realizes that the executive has bypassed the lawyer's review of the project and is prepared to allow the release without the normally required lawyer's approval. The lawyer's duty to protect the company's reputation, abide by their policies and avoid potential negative publicity from the release will require further inquiry and analysis. This will delay the release and impact holiday sales targets. If the lawyer's only recourse is to the impacted executive, the result may be different than if the lawyer is able to access support from the general counsel to create a broader review of the economics of distribution versus implementing company values. Because the embedded lawyer understands the business issues and pressures affecting the division's executives, the lawyer can best represent the division's priorities to the general counsel while also serving as a mediator of the company's values as expressed by the general counsel. In other words, the business executive usually has a primary, short-term economic focus, whereas the general counsel has a longer-term reputational and corporate policy and values focus. The embedded lawyer can translate, listen and mediate these differences for the best outcome for all parties. Thus, the company, the true client, can meet its long- and short-term economic and values objectives.

The embedded lawyer/general counsel relationship must be built from shared values and understanding of foundational company goals. In the example just given, the company's family-friendly reputation will take precedence but must be pragmatically applied. Prior communication on these topics is essential as the base from which the embedded lawyer will approach the general counsel on any particular, immediate concern. Being able to reference this reputational priority based upon the established company goal and prior dialogue will create the context that allows the embedded lawyer to depersonalize the immediate issue from a particular product or manager/executive and allow for a pragmatic examination of the business opportunity versus the reputational goals.

Also in the example, because the embedded lawyer is the gateway that the transaction must pass through (i.e. the manufacture and distribution contracts to complete the doll), the lawyer can bring the general counsel and/or public relations executives into the discussion of short-term profit versus the long-term reputational goals of the company. By candidly addressing each party's motivations and interests in the context of the lawyer's and company's "family-friendly" values, all interested parties can address the effects of the business decision in a manner that will allow the final decision to integrate values into the business/profit motivations of the manager. The lawyer's duty to the values of the company (and his own values of honesty and candor as a spokesman for the company's "family-friendly" values) will allow the manager to see the manager's self-interest in the light of the reputational value of the company rather than solely in terms of his production/profit goals and his own short-term compensation for getting the inappropriate doll to the market. Additionally, the lawyer can serve as an exemplar of higher values and trigger such values in the manager. These conversations can be inclusive, not confrontational.

The next chapter describes the importance of establishing objectives of representation between the lawyer and client. The embedded lawyer should clarify the overriding consideration of protecting the company's family-friendly reputation and the general counsel's role as final arbiter in order to avoid surprises to the executives who are pushing to implement their plans within the construct of specific relevant business and legal constraints. The lawyer's ability to candidly advise the executives and the company will benefit from early explanation of the primacy of the company, not the executive, as the client.

Note

1 American Bar Association Center for Professional Responsibility, *Model Rules of Professional Conduct* (Chicago, IL: American Bar Association, 2019).

Chapter 7

Objectives of representation

A lawyer must understand the client's objectives in order to properly advise the client of the risks and potential outcomes of the transaction or litigation that is the subject of the engagement. This chapter will describe the counseling methods that will enable the client and lawyer to come together to prepare their objectives of representation and describe the path that will follow. This predictive plan should set forth estimates of time, costs and decision points based upon the client's goals and the lawyer's experience in similar matters. Model Rule 1.4, *Communication*, and Model Rule 1.3, *Diligence*,[1] require the lawyer to update the plan and stay in contact with the client at all stages of the engagement. A thoughtful plan is necessary to implement the legal, business and ethical strategies of the lawyer and the client as well as avoid surprises and tensions on values issues.

Model Rule 1.2, *Scope of Representation and Allocation of Authority Between Client and Lawyer*, generally says that although the client has "ultimate authority to determine the purposes to be served by legal representation," the lawyer should control "the means to accomplish the client's objectives . . . with respect to technical, legal and tactical matters." The client should control the expenses to be incurred and consider the effect of same on third persons who might be adversely affected.

Another element to be considered in setting the objectives of representation is Model Rule 2.1, *Advisor*,[2] which states:

> In representing a client, a lawyer shall exercise independent professional judgment and render candid advice. In rendering advice, a lawyer may refer not only to law but to other

considerations such as moral, economic, social and political factors, that may be relevant to a client's situation.

In summary, the objectives of representation are a road map jointly arrived at by the client and lawyer at the commencement of the engagement. The client is responsible for the goals of representation and the lawyer is responsible for the means to accomplish them. The client should be able to take advantage of the experience of the lawyer to predict costs, decision points, time and risks.

It is important that the client and lawyer use the process of setting the objectives of representation to discuss and align their respective values to avoid unexpected conflicts at various decision points in the representation. Anticipating ethical and values conflicts will allow for reasoned discussion and finding alignment between client and lawyer.

The ethical rules are clear that the lawyer is expected and entitled to give their candid advice, not just their legal advice. Also, Model Rule 1.2(c)[3] makes clear that representing a client does not constitute approval of the client's views or activities. The lawyer must remain independent of the client to provide objective advice and need not agree or identify with all aspects of the client's activities unrelated to the representation. The independent legal and values-oriented advice of the lawyer should be communicated with the client in the course of the representation and especially in setting the objectives of representation in order to avoid future unexpected values and/or ethical conflicts.

Model Rule 1.18[4] defines duties to a prospective client. The first contact between a person seeking legal representation and a lawyer makes that person a "prospective client." The ethical rules and case law require that the burden of defining that relationship rests on the lawyer. This is the first opportunity the lawyer has to establish compatibility. Model Rule 1.16 suggests that a lawyer may decline representation or, for enumerated circumstances, withdraw from representation. The right of a lawyer to decline representation for reasons other than competence is an important aspect of values-oriented practice but must be carefully considered and exercised.

Although a lawyer is not expected to share all of the political and social opinions of their clients, there are certain instances where the divide between the lawyer's personal, moral objectives and that of the client diverge so sharply that it is best to decline representation rather than pursue a client's objectives that are repugnant to the lawyer. Clearly, the client that demands that the lawyer falsify documents or lie on the client's behalf should be terminated as a client. However, there are less clear instances when a client's and lawyer's objectives are so divergent that good-faith representation is not feasible. Strong contrary views between lawyer and client on such issues as reproductive rights, for example, (if such issues are relevant to the desired representation) may be cause for declining representation. Certain tactics desired by the client, while within the bounds of propriety, may also be cause for denial of representation. An example would be as follows: in a product liability lawsuit involving allegations of defective manufacture of intrauterine devices, the defendant company adopted the strategy of deposing all plaintiffs and inquiring in explicit detail about sexual promiscuity, personal hygiene and evidence of mental illness. The objective was to make the deposition process so uncomfortable for the plaintiffs that settlements would be easier for the defendant company. A lawyer tasked by the firm to participate on behalf of the defendant company would, in my opinion, be justified in declining such assignment even though other lawyers in their firm may choose to proceed.

In addition to withdrawing from participating in future activities on the case, the lawyer could meet with senior leaders in the firm and describe the visceral reaction he had to the tactics of the litigation as an assault on decency and the value of compassion for the plaintiffs. The lawyer could explain that if he had such a negative reaction, the tactics may ultimately harm the defendant client more than help the case. The senior firm leaders may be moved to reassess the strategy and consult with the client to change tactics in a way that is more in line with the law firm's and the client's universally held values. Strategies of "win at all costs" often can have a significant backlash effect on judges, juries and the public, and such tactics can lead to undesirable outcomes. Whatever the tactical effectiveness of this strategy, the expression of the lawyer's individual values can be a powerful message that may cause others to reconsider their actions and express their own values more robustly. Speaking up when you know what's right can convey a powerful message.

As with many relationships, the first stages of the attorney-client relationship will set the tone for the entire relationship. The attorney must use the first encounter to fully understand the client's objectives in order to determine whether they can render effective representation consistent with their skills, experience and values. An experienced practitioner should seek to listen carefully to the client's perception of what the attorney should accomplish on their behalf. At the same time, the attorney should clearly set forth time, costs and likely critical decisions that the attorney and client will be making in order to accomplish the client's objectives. At this early stage the attorney should seek to explain their approach to the problem solving, their assumption of risks and their values as they relate to all who will be affected by the outcomes sought by the client.

Assessing compatibility on key objectives of the client and the manner in which the lawyer will conduct the representation should establish the client's assent to the scope of representation and allocation of authority described earlier in Model Rule 1.2[5] (the client determines the purpose of the legal representation and the lawyer determines the means to accomplish the client's objectives with respect to technical, legal and tactical matters). This principle is importantly impacted by Model Rule 2.1, where the attorney as advisor shall exercise independent professional judgment and render candid advice. A lawyer does not need to adopt or conform to aspects or activity of the client that is not related to the legal representation. For example, the lawyer's or client's political views, social views or economic circumstances may be irrelevant to the legal representation sought or rendered. Mutual respect and nonjudgmental conduct on these unrelated factors will aid both parties in aligning business, ethics and values to accomplish the desired outcomes of the representation but may not be a requirement.

Simply stated, the values of candor and honesty by the lawyer and the openness of the client to accept and consider such candid and independent professional judgment are the cornerstone of setting the road map of the objectives of representation.

The early agreement of a plan and budget to accomplish the client's objectives with lawyer and client being "equal partners" in their respective roles will allow both sides to make clear decisions and anticipate and adjust for any unforeseen issues that may arise in the course of their collective efforts to achieve the client's objectives.

Lawyers and legal processes are often seen by clients as tools to accomplish their economic goals. The following example demonstrates the implementation of the ethical rules discussed in this chapter, allowing the client to best utilize the lawyer's experience and legal skill to shape the objectives of representation for an outcome consistent with the lawyer's and client's objectives.

A real estate developer has acquired a city block in an underserved neighborhood in order to redevelop the block into mixed-use retail and housing. The redevelopment envisioned by the developer will require eviction of the current tenants, demolition of the existing buildings, rezoning and building variances. The client asks the lawyer to represent them in the process with the expectation that the attorney will file eviction proceedings to clear the property. The lawyer advises the client that based upon the lawyer's experiences, the evictions will create a political backlash and will make rezoning or building variances problematic (if they could be achieved at all). The lawyer advises the client that zoning benefits for mixed-income housing also have tax benefits that will mitigate the developer's costs and may allow for keeping some existing tenants. This will provide stability for the project, maintain the community and provide local support for rezoning and redevelopment. Assuming the client has values similar to the lawyer regarding maintaining community and mitigating gentrification and displacement, the lawyer's willingness to express their values and experience may provide a more desirable outcome than wholesale eviction proceedings. Perhaps, however, the client has investors or lenders who don't share these values. The lawyer can help the client (with the aid of other experts such as political consultants, appraisers or tax specialists) by providing data and relevant comparable examples of the increase in value, efficiency of execution and reduction of legal risk that are additional factors for adopting the lawyer's recommendations and accomplishing both improved execution and economics as well as a values-oriented outcome.

If the client declines the lawyer's advice and chooses to proceed with evictions and demolition, the lawyer should explore alternatives that keep the core of the client's strategy but add mitigation for the displaced tenants in a manner that benefits the client's business objectives but is consistent with the lawyer's best advice and socially responsible values. In the absence of a pragmatic solution that satisfies both the lawyer and the client, the lawyer may decide to decline the representation so that the client can find a lawyer with more compatible goals and values.

In order to provide effective business and values-oriented advice, the lawyer should access financial modeling resources that explain the tax benefits of low-income housing tax credits on the overall financial performance of the project. Such credits can facilitate keeping some existing tenants in the neighborhood, thus serving the social value of not displacing long-time neighbors or local merchants and services.

Also, the lawyer can encourage engagement of political consultants or, with the client's involvement, meet with local planning officials and elected representatives to help quantify the zoning bonuses or variances available as a result of maintaining a cohesive tenant base, in addition to the new development potential of revised entitlements. Appealing to the client with hard evidence and data on the economic incentives reinforces the values of community and compassion for existing families and maintenance of their homes and neighborhoods, which will serve and enhance the values of both the lawyer and the client.

Notes

1 American Bar Association Center for Professional Responsibility, *Model Rules of Professional Conduct* (Chicago, IL: American Bar Association, 2019).
2 American Bar Association Center for Professional Responsibility, *Model Rules of Professional Conduct* (Chicago, IL: American Bar Association, 2019).
3 American Bar Association Center for Professional Responsibility, *Model Rules of Professional Conduct* (Chicago, IL: American Bar Association, 2019).
4 American Bar Association Center for Professional Responsibility, *Model Rules of Professional Conduct* (Chicago, IL: American Bar Association, 2019).
5 American Bar Association Center for Professional Responsibility, *Model Rules of Professional Conduct* (Chicago, IL: American Bar Association, 2019).

Chapter 8

Counseling/decision making in the attorney-client relationship

Before a lawyer encounters the adversary in litigation or the counterparty in a transaction, they must first establish the objectives of representation with the client, based upon careful listening to the client's needs and objectives and upon the application of the lawyer's skill and experience to the facts and the law. They must also maintain clear and updated communication with the client as the circumstances of the representation change through negotiation activity and decisions of third parties at various points in the evolution of the litigation and/or the transaction.

The most effective method for the creation and implementation of the objectives of representation is counseling.

Counseling allows the lawyer to provide independent, candid advice in addition to their more public-facing role as an advocate or negotiator on behalf of the client. As an advocate, the lawyer often expresses the outer range of objectives or demands the client wishes to put forth in early stages of litigation or negotiation. The two roles of advocate and counselor often create tension with values choices, as the client may wish the lawyer to express a position or course of conduct which may be technically permissible under existing law though inconsistent with its underlying purpose or spirit. Recall here the divorcing husband who wanted to pursue an insincere attempt for child custody to extract lower child support obligations. Model Rule 2.1[1] requires lawyers in an advisory role to "exercise independent professional judgment and render candid advice." A lawyer must establish their independence under all circumstances in order to preserve a sufficient detachment from their client's interest in order to provide sound and objective evaluation of the ethics, values and business efficacy of the client's choices at all stages of the litigation or transaction. Clients want skill, experience and judgment from lawyers. The lawyer's ability to gain the client's confidence that the lawyer has sufficient skill, experience and judgment and/or access to those qualities among the lawyer's colleagues will go a long way in giving the lawyer the standing to provide advice that the client will accept and enact.

The comment to Model Rule 2.1[2] clearly demonstrates the meaning of the twin requirements of independence and candor:

> Legal advice often involves unpleasant facts and alternatives that a client may be disinclined to confront. In presenting advice, a lawyer endeavors to sustain the client's morale and may put advice in as acceptable a form as honesty permits. However, a lawyer should not be deterred from giving candid advice by the prospect that the advice will be unpalatable to the client.

As described in Chapter 3, ethical and values conflicts are a constant and not occasional.[3] Therefore, a lawyer should establish a counseling relationship that has candor and independence as a foundation so that "hard truths" can be freely discussed in both the anticipatory design of the objectives of representation and the implementation of those objectives through the counseling relationship.

The ethical and values-oriented aspects of client counseling are generally classified in three modes: (i) client-oriented, (ii) collaborative, and (iii) lawyer-centered. These forms of counseling are not rigid and may overlap on certain issues. Legal advice couched in narrow technical terms is normally not valuable to the client because practical considerations such as costs or effects on other people typically require exploration of the ethical rules and the values of both the client and the lawyer.

Client-oriented counseling prioritizes client autonomy. Collaborative counseling makes lawyers and clients partners in goal setting and problem solving, and it promotes joint responsibility for ethical and values-oriented outcomes. Lawyer-centered counseling acknowledges the lawyer's greater responsibility for the consequences of professional advice (often regarding areas of specialization such as tax advice).

In the prior example of the real estate developer hiring the lawyer for evictions to enable demolition and redevelopment of existing housing, the client-oriented approach would involve an explanation of the mechanics and costs of eviction proceedings. Collaborative counseling would provide alternatives that utilize the lawyer's experience and values to help accomplish a more complete and nuanced result. Lawyer-centered counseling would examine the social implications of gentrification while applying beneficial legal incentives such as zoning bonuses and mixed-income housing tax credits.

Client-oriented counseling is an analytical model that is, in some respects, a reaction to the role of the professional in the lawyer-centered model as an expert decision maker. The medical profession has been criticized as too often failing to provide full explanations of treatment options or details necessary for a patient to predict outcomes and effects of various alternative therapies or treatments. Many patients have experienced an appointment where the doctor tells the patient the next treatment without the opportunity to discuss patient concerns or fears and to respond as equals deciding treatments. Many lawyers, especially in technical areas of the law, take a similar approach. The lawyer-centered model could take into account the lawyer's values orientation, but without client input the lawyer never hears the client's values and therefore may offer advice that is inconsistent with those values.

Lawyer-centered counseling has particular merit in representation of economically disadvantaged clients or clients whose impairments stem from youth, peer pressure or psychological trauma (such as divorce). Such clients might not feel the autonomy afforded to clients more comfortable with interactions with professionals and technical decision making. Client-oriented counseling, on the other end of the spectrum, may involve sophisticated clients who seek to have the lawyer give purely technical legal advice and implement the client's instructions without full exploration of the ethics, values or impact to third parties. The risk of the client-centered approach is that the lawyer may follow directions that reflect values that may harm third parties or the client themselves. In situations where clients are under stress of impairments or focused on short-term outcomes or profits (e.g. personal injury settlements or eviction proceedings), they may be poorly situated to take a long-term view of their interests or live up to their values. Therefore, a client's instructions to take certain legal actions may benefit from the independent, candid advice regarding values that balance their long-term interests with short-term goals.

A definition of the professional role that encourages deference to clients' current preferences may poorly serve their ultimate interests. Importantly, if the client's understanding of the role of the lawyer as having independent perspective on values choices is not established at the beginning of the lawyer/client relationship, the expression of candid ethical and values advice may be seen as an obstacle rather than as enhancing the client's objectives and satisfying the lawyer's personal and professional responsibilities.[4]

The collaborative approach follows the plan outlined in Chapter 7 of cooperatively preparing an objectives-of-representation road map between the client and lawyer, each listening and contributing to a viable partnership. This will accomplish the client's objectives in a manner that considers ethics, values and business goals and that provides for communication of alternative choices as the litigation or transaction evolves and affects multiple parties and decisions. Collaborative frameworks recognize that one of the most valuable contributions is to engage and enlarge the client's values vision and to encourage decisions that express the parties' highest principles. The collaborative approach favors dialogue that is a genuine two-way street.[5]

Lawyers should be prepared to explore the client's values and emotions as part of counseling. Choices of alternatives very often require exploration of psychological and emotional needs as well as financial outcomes. Although Model Rule 2.1[6] gives lawyers leeway in giving values and ethical advice, comments to that rule make clear that the lawyer, as counselor, should also recognize areas that are the domain of other professions, such as psychiatry and social work for family matters or accounting and financial specialists for business matters.

In our prior real estate redevelopment example, perhaps an introduction to a political consultant or community organization will broaden the client's perspective and understanding of their legal, political and social objectives.

The collaborative approach best serves the goal of finding a balance between ethics, business and values. Both the client and the attorney put their objectives on the table in each of these areas so that any compromise of business objectives or tactical decisions (within ethical limits) is fully and openly discussed, the benefits and burdens of the potential decision are analyzed and both the attorney and the client mutually agree to and support the course of action arrived at after the candid assessment. If the lawyer (in the lawyer-centered approach) or the client instructions to counsel (in the client-oriented model) were followed without the informed input and agreement of the other party, there would be lingering resentment and uncertainty and potentially lack of trust in or commitment to future decisions in the representation.

The client's knowledge of their business and their duties to their constituents (shareholders, family members, employees, regulators) can often cause adherence to a narrow course of action that the lawyer as an independent professional—experienced in similar situations and aware of all potential legal and social ramifications of a certain course of action—may be able to temper or adjust in a way that creates better long-term outcomes for the client consistent with the objectives of representation. The value of this independent professional judgment can only be realized through both parties' mutual respect for honest, candid communication and shared values of fairness and responsibility.

The risk of collaborative counseling in the era of increased legal specialization and business pressure on the lawyer to produce clients, hours and billings is that the lawyer may lose some of the necessary detachment and perspective necessary to ensure the proper balance of business and ethics. This issue is addressed in several examples in this book, but most clearly in Chapter 10, regarding Enron. Although the lawyer may avoid charges of fraud for aggressive legal advice, the lawyer should provide a safe distance for the lawyer and the client from the appearance of enabling fraud. Independence and detachment can be reconciled with collaboration, but the source of that independence is a firm grounding in the values discussed herein.

Legal principles and ethical rules have evolved over time to respond to financial fraud and crime. Model Rule 1.6 regarding client confidentiality begins with a qualification that the lawyer may disclose confidential information to the extent the lawyer reasonably believes such disclosure is necessary to prevent harm. In 2003, in response to Enron and other frauds, the model rules were amended. While Model Rule Section 1.6(b)(1) previously allowed the disclosure "to prevent reasonably certain death or substantial bodily harm," the 2003 amendments added Sections 1.6(b)(2) and (3) to expand the circumstances of allowed disclosure to "substantial injury to the financial interests or property of another" for future crimes or frauds ([b][2]) or past crimes or frauds ([b][3]). As with all such rules, there are narrow guidelines that must be satisfied to comply. Attorneys representing clients before the Securities and Exchange Commission (SEC) have certain rights and responsibilities under rules promulgated by the SEC.

Notes

1 American Bar Association Center for Professional Responsibility, *Model Rules of Professional Conduct* (Chicago, IL: American Bar Association, 2019).
2 American Bar Association Center for Professional Responsibility, *Model Rules of Professional Conduct* (Chicago, IL: American Bar Association, 2019).
3 Mary C. Gentile, *Giving Voice to Values: How to Speak Your Mind When You Know What's Right* (New Haven, CT: Yale University Press, 2010), 72. www.GivingVoiceToValues.org/
4 Deborah L. Rhode, David Luban, Scott L. Cummings, and Nora F. Engstrom, *Legal Profession*, 7th ed. (St. Paul, MN: West Academic, 2016), 491.
5 Deborah L. Rhode, David Luban, Scott L. Cummings, and Nora F. Engstrom, *Legal Profession*, 7th ed. (St. Paul, MN: West Academic, 2016), 492.
6 American Bar Association Center for Professional Responsibility, *Model Rules of Professional Conduct* (Chicago, IL: American Bar Association, 2019).

Chapter 9

Confidentiality and conflicts of interest

The Duty of Confidentiality as embodied in Model Rule 1.6 and the multiple Rules governing Conflicts of Interest (Rule 1.7, 1.8, 1.9, 1.10 and 1.11)[1] are complex and porous but sit squarely at the intersection of ethics and business in the lawyer's practice. The issues and actions required to meet the business objectives of the client, the development of the lawyer's career (often with conflict between the lawyer's colleagues) and the ethical rules bring into sharp relief the implementation of the lawyer's values and relationships. It is beyond the scope of this book to explore all the intricacies of these ethical rules or the multiple challenges caused by observance of the varied interests impacted by the Duty of Confidentiality and/or the rules regarding conflicts of interest.

The purpose of the rules at issue is to create trust and a safe space between client and lawyer. In order to best meet the objectives of representation, the lawyer and client must feel that the facts of the situation and the goals of the client can be freely and candidly discussed at all stages of the representation. The trust between lawyer and client is made possible by the protections of confidentiality, which are enabled and reinforced by these rules.

As a useful aside, the Duty of Confidentiality, an ethical rule, is not exactly the same as the Attorney-Client Privilege, a rule of evidence, but the two reinforce each other to protect client confidences. The Duty of Confidentiality relates to the lawyer's obligation to the client to preserve and protect the information shared with the client in furtherance of a specific legal representation. This ethical duty is enforced by sanctions against the lawyer for violations. The Attorney-Client Privilege is a rule of evidence, enforced during judicial proceedings by the Court, to protect communication between attorneys and clients in specific legal representations. There are specific technical characteristics and numerous exceptions to the rules. However, it may be useful to test the application of the rules not by whether the client is harmed by the release of the relevant information (although there may be other significant consequences to be considered) but rather whether such release is an assault on the sanctity of the Duty to Protect Client Information in a way that undermines the system of trust these rules were intended to create.

The preamble to the Model Rules contains the following sentence:

> Virtually all difficult ethical problems arise from a conflict between a lawyer's responsibilities to the clients, the legal system and the lawyer's own interests in remaining an ethical person while earning a satisfactory living.[2]

Once the client is convinced that the lawyer is loyal to the goals of the client and will protect the client's information, the client will form the counseling bond necessary to utilize the lawyer's knowledge, skill and experience. It is at that point that the lawyer can candidly offer their advice and express their values to confirm the alignment necessary to create a plan of representation and respond to changes. In other words, client confidentiality is a pillar of trust between client and lawyer. See Chapter 11 regarding the importance of trust. The sharing of sensitive (often damaging) information creates a necessary vulnerability between the client and the lawyer. The confidence that the client has that the lawyer will protect the client's confidence can define the relationship.

The rules regarding conflicts of interest are similarly designed to reinforce the lawyer's duty of loyalty and client confidence. The evolution of specialization of law practices has created lawyers who represent clients based upon their technical (expert) skill, not as a general practitioner on a wide range of the client's legal problems. Therefore a premium is placed on finding a lawyer with previous involvement on the subject in which the client seeks expert representation. The promotion of the lawyer's expertise, including prior representations of clients in similar industries or with similar issues in transactions, regulatory or litigation proceedings, is an important business development tool. As a lawyer builds their practice, the client is interested in finding out the lawyer's prior experience with others in their field and, often, competitors. It is a common business development tactic to seek to impress the larger, more prestigious counterparty to a transaction in hopes of gaining a new larger and more profitable client. The lawyer's individual interest in building their practice can be a stress factor on the duties the lawyer has to their existing client. The value of loyalty and diligence to the current client in the face of the potential for a more prestigious future client can challenge the lawyer's balance between ethics, business and values.

The role of the large law firm in conflicts of interest is another challenging factor. Advancement in a large firm is often influenced by the lawyer's ability to bring in representations of prestigious and profitable clients. An additional issue is the imputation of conflicts described in Model Rule 1.10,[3] where the conflict of one lawyer may disqualify the entire firm of hundreds of lawyers from representing a client. There is probably no more intense business/ethical conversation in a large law firm than the clearance of conflicts, the attribution of client origination or the interpretation of the exceptions to the conflicts rules in the circumstances of an imputed conflict. The expression of the values of the lawyer (including client diligence and loyalty) and the expression of the values of the law firm in resolving these issues are important markers of the integrity of the lawyer and the firm.

The guiding value in issues of confidentiality and conflict of interest issues should be to balance business interests and ethical compliance with the value of diligence to meeting the current client's objectives of representation and preserving the trust of this client and future clients in the loyalty expected by the client from the lawyer.

The conflict of interest rules, attorney/client privilege and the duty of confidentiality are not always used by lawyers and clients as a method to preserve the trust between a lawyer and client. The misuse of these rules for strategic advantage or as an offensive tactic is a common abuse of the rules.

The conflict of interest rules are frequently strategically asserted in litigation to disqualify the counsel of the client's choosing, or at least embarrass chosen counsel, delay proceedings or increase the costs of litigation. The attorney-client privilege and the duty of confidentiality may be used to shield socially undesirable information or harmful products from discovery at the cost of violating a lawyer's values or risking harm to third parties that would be safer if the information were public. The evaluation of the ethics rules and the laws governing attorneys' conduct have recognized the need for balance of the rights of third parties affected by litigation or transactions with the sanctity of attorney-client confidentiality and attorney loyalty. In each instance where the communication or loyalty of the attorney is at issue, business and ethical rules and exceptions to the rules require "the exercise of sensitive professional and moral judgment guided by the basic principles underlying the rules."[4]

As earlier described, the first step of effective representation is through understanding of the facts and application of those facts to relevant law. Before a lawyer can competently render legal or business advice, they will need to gain agreement with the client on the facts of a situation and a fair concurrence with the client on available risks and likely outcomes. Until the client and lawyer are on the same page about the facts and the client is informed about relevant legal rules, the lawyer may antagonize the client or other parties by revealing information or advocating solutions that will cause opposition to the lawyer's business or legal advice or reveal failure to fully integrate the lawyer's and client's values as applied to the situation at hand. In order to move in concert with the client on business, legal and values actions pursuant to the objectives of representation, it is essential that the lawyer command a confident, accurate understanding and expression of the relevant law and facts.

If the lawyer determines that the client's understanding of the facts and law still creates a conflict in values, the ethical rules of client confidentiality restricts the lawyer's ability to proceed, as the choice of business alternatives are the client's prerogative. For example, assume you represent the owner of trailer parks that lease to senior citizens (i.e. people over the age of 55). Most of the residents live on Social Security or fixed incomes. The client has maintained a long-standing but undocumented policy of allowing his tenants to convert their tenancy to ownership of their dwelling unit if they maintain consistent payment history and cause no social disruption after five years of occupancy. The residents do not have a legal right to insist on the conversion to ownership and are subject to eviction as month-to-month tenants. The client has the power to affix the mobile homes to the property, thus qualifying the tenants for much lower-cost long-term financing at the same monthly mortgage payment as the rent and a capital gain to the client. The client informs you that he no longer wants to continue this practice and wants to begin evictions to start leasing to a wider group of tenants at higher rents. You are convinced (and believe you can effectively demonstrate) that this will put the client at greater economic risk than the steady income and capital gains from the prior strategy, but your greatest concern is the disruption and hardship visited on the elderly tenants who would need to find new housing.

Your knowledge of the consistency and mechanics of the past practices could be valuable to tenant groups who may seek political or legal help in resisting the evictions or pursuing other financial alternatives to present to the landlord. You are precluded by ethical constraints of client confidentiality from aiding the tenants without the client's consent. Your only alternatives are to (1) resign in protest, or (2) fully research and explain legal and business alternatives that have equal or greater business outcomes and are consistent with the value of compassion and thereby conform to shared values.

Assuming the client has made the business decision to pursue a strategy of evictions under the impression that he will be free of risk and therefore able to implement the rental model over the sale model, the lawyer should still seek to explain the efficacy of continuing the past practice in economic, legal and values terms. First, the lawyer can help the client best understand the various economic alternatives available from the sale model, including self-financing, contracts for deeds or other long-term sale arrangements that the client may not have contemplated but which may yield superior economics and lower risk than the lease model. In this circumstance, the lawyer uses his experience to counsel the client on unexplored alternatives by deconstructing and reexamining the business model. Second, the lawyer can explain the political and litigation risks that could ensue if the client alters past practices that may have been reasonably relied upon by the tenants or could be negatively perceived by elected officials with authority over the property. A fair risk analysis is always appropriate when there is a significant change of conditions and practices. The reputational risk to the client should also be discussed, and the optics of these changes to the tenants' situations may put the client in an undesirable public light. Finally, the lawyer can, and should, express his values of maintaining the community, following through on implicit practices and compassionately treating the tenants who have "invested" in their homes reasonably relying on past practices.

Once the client's objectives of representation have been fully advanced, the lawyer and client can consider where their respective business goals are best served and proceed honestly and candidly from the foundation of trust established by adherence to their shared values.

There are several practical guidelines that attorneys should consider in the client confidentiality and conflict of interest context. First, don't discuss cases, clients or deals with friends or people who aren't directly involved (even those in your firm). Second, try not to explain your expertise by reference to identifiable clients or deals you have been or are involved in. In this regard, hypothetical or law-based discussions may explain your experience but care should be taken to not use examples that would reveal specific clients or transactions.

Third, be careful of discussions in public places (elevators, airplanes, etc.). No matter how remote you believe you may be, coincidences abound and you never know who may recognize, and utilize, information you reveal. Fourth, there are ample examples of careless use of email or social media to inadvertently reveal confidential information. Always pause and re-read messages (including all recipients' addresses) before hitting "send." Revealing details of your personal life and travel may also reveal information that can be linked to clients, cases or deals which could be disadvantageous to your client or firm.

Trying to build your reputation through reference to the confidences or prestige of your clients or your involvement in specific cases or deals may have unintended negative impact on potential clients who may be concerned that you would reveal details of their business indiscriminately or for purposes other than the advancement of their goals.

Notes

1 American Bar Association Center for Professional Responsibility, *Model Rules of Professional Conduct* (Chicago, IL: American Bar Association, 2019).

2 American Bar Association Center for Professional Responsibility, *Model Rules of Professional Conduct* (Chicago, IL: American Bar Association, 2019).

3 American Bar Association Center for Professional Responsibility, *Model Rules of Professional Conduct* (Chicago, IL: American Bar Association, 2019).

4 American Bar Association Center for Professional Responsibility, *Model Rules of Professional Conduct* (Chicago, IL: American Bar Association, 2019), 1–5.

Chapter 10

Enron

Lessons in corruption of values

As an introduction to this chapter on Enron, it may be helpful to understand my perspective and background, which resulted in the analysis described in this chapter. My law firm was appointed as Debtor's Counsel to the Enron Estate by the Bankruptcy Court with jurisdiction over the Chapter 11 case. My firm had no involvement or representation of Enron prior to the restructuring through the Bankruptcy Code proceedings. Because many Enron creditors were directly involved in the structuring and financial matters related to the "off balance sheet" financings (accomplished through the use of special purpose entities as described in more detail in this chapter), the court appointed a Chief Restructuring Officer for the Debtor's Estate. It appointed my law firm as its counsel, and it urged the Bankruptcy Court to appoint an Examiner, pursuant to the Bankruptcy Code, to investigate the "off balance sheet" financings. Typically, such investigations would be conducted by the court-appointed Official Creditors' Committee, but it was felt that, due to the deep involvement of many of the major creditors in the subject transactions, an independent Examiner would provide a more thorough review.

Although my practice specialties did not include bankruptcy, I had worked on many financings and restructurings as a corporate attorney. As such, I assumed the role within the Debtor's Counsel team of gaining an understanding of the facts and law surrounding the "off balance sheet" financing and coordinating with the Debtor's Estate, the Creditors' Committee and, most particularly, the Examiner to prepare the Examiner's Report to the court. Armed with the conclusions of that Report, our firm represented the Debtor's Estate, acting through the court-appointed Restructuring Officer, in negotiating settlements with the counterparties to the special purpose entities and gaining court approval for those settlements.

This role caused me to interact with many of the participants to the "off balance sheet" financings, including Enron finance executives, in-house and outside counsel for Enron and creditors and the Creditors' Counsel. I was also a liaison between the Debtor's Estate and the Examiner and his counsel and helped supply some of the information included in the Examiner's Report.

The observations of this chapter come from the vantage point I gained in the role just described. My objective is to only discuss information that is publicly available and not utilize any privileged or confidential information gained through attorney-client communication. Given the transparency expected by the bankruptcy process, I believe the descriptions herein satisfy my objectives.

The circumstances of the collapse of Enron, a major American corporation, involve complex factors that were investigated from many vantage points. The question that was often propounded during those investigations was "Where were the lawyers?" This chapter presents an overview of that failure as a cautionary tale, with a particular focus on the role of the lawyers and the corruption of values that occurred.

Background

In the Fall of 2001, Enron Corporation—a global energy company based in Houston, TX—was the seventh-largest company on the New York Stock Exchange. The company's stock price rose consistently based upon the perception that it was well managed, in a growing industry and had consistently rising quarterly and annual earnings. The collapse of the company through an unplanned bankruptcy was a shock to global markets and led to government inquiries as to the cause of the failure.

A major cause of the collapse was credited to the use of suspect accounting practices and financial reports that misstated debt as income, manipulated through "off balance sheet" partnerships that timed reported income to artificially show consistent financial growth and stability.

As described in more detail here, Enron and its lawyers sought to employ a financing technique often referred to as "structured finance," based upon an accounting rule that allowed for the sale of a partial interest in an asset to a partnership where the seller retained an interest and a measure of continued management. Because the transaction appeared to meet certain accounting rules and criteria, the accountants would record the value of the sold interest as income based upon the risk assumed by the buyer/financier of ownership of the asset. However, Enron—and its counsel—disguised the true nature of the financing by a disguised guarantee and a bogus legal opinion which was not fully disclosed in the financing statements. Additionally, the company lawyers and the outside counsel responsible for the regulatory reporting of financial conditions obscured the true nature of the transactions.

The conflicts of interest by the in-house lawyers, the outside law firms and many of the individual lawyers (including senior legal leaders of the company and senior partners at the outside law firms) represent stark lessons in the factors that can lead to the corruption of values of lawyers in the representation of opportunists in the corporate context. The authority the Enron corporate finance executives had on the business motives of the lawyers was also a corrupting influence.

In the Enron case, the Bankruptcy Court appointed an independent Examiner to review the role of the parties to the accounting-driven "off balance sheet" transactions, which were integral to the financial manipulations and misstatements of the financial conditions of Enron. As explained in the Examiner's Report,[1] the misuse and failure to expose such misuse of the attorney's "true sale" opinion allowed Enron to record debt as earnings. This was the flawed foundation of the financial reporting. Part of the Examiner's Report chronicled the role of the in-house and primary outside law firms and lawyers who oversaw the flawed transactions. Many lawyers who worked for outside law firms on various financing transactions were subsequently hired by Enron as in-house counsel (on compensation arrangements greater than their prior employment). Those same lawyers then worked on the extensions, amendments and increased funding for Enron, utilizing their knowledge of the funding financiers as well as their relationships with their former colleagues who now were the lawyers for the financiers in the amended, extended transactions. The "second stage" amendments and extensions added billions of dollars of risk to the company, its shareholders, pensioners, employees and the shareholders and employees of the financiers. The circle and cycle of complicity among the executives, bankers and lawyers was a material factor in enabling the fraud to continue. As described in the Examiner's report,[2] several younger partners at Enron's primary law firm brought their concern to senior management. The managing partner of the firm and Enron's general counsel (a former partner at the same firm) both had built their status and careers around this one very important client. Neither wanted to be responsible for exposing the flawed legal opinion, thereby allowing for a catastrophic corporate failure which affected thousands of investors, employees, utility rate payers and many charitable and community organizations in Houston and beyond.

The Examiner's Report describes the consequences of silencing the young partners who sounded the alarm. They were subjected to enormous career pressure and the rules of attorney confidentiality to not take their complaints outside their firm and the company. We all can hope that, with the benefit of hindsight, we would have been more courageous.

Once the Enron General Counsel and the Managing Partner of Enron's primary outside counsel advised the younger partners that further actions on their concerns were not to be forthcoming (i.e. the misused legal opinions would be left at status quo), there was, under the then prevailing ethical rules of Attorney-Client Privilege, no further recourse available other than quitting their jobs in protest. Enron and other corporate frauds of that era (e.g. WorldCom) spurred a congressional inquiry and legislation (the Sarbanes-Oxley Act) which created enhanced reporting outlets for attorneys in this situation. It is beyond the scope of this book to explain those legal remedies.

As described in the previous chapter, rigorous exploration of the facts and law is an essential component of effective advocacy of values. The post Enron rules largely direct the lawyer to report the unacceptable conduct "up the chain," as was done at Enron, and then to independent members of the governing Board of Directors. In Enron, an employee made a similar complaint regarding accounting practices to the Board and it was largely ignored and whitewashed. If the opportunists in control of decision making want to cover up the bad conduct, it is often done through falsification or misrepresentation of the facts to the larger internal audience. In Enron, the lawyers were told that the misuse of the "true sale opinions" was an accounting issue, not a legal issue (i.e. not in their purview). One can only speculate the outcome of a very well-researched memorandum that refuted the abuse of the "true sale opinions," the risks to the company and the law firm of these tactics and exposure of the opportunists and their activities. Would that have caused a different result?

Given the pace of the deceit and profits created by these tactics, my observation is that the likely result would be that the career of the author of such a memo would be seriously jeopardized, but no action would be taken to correct the abuse. Whistleblower protections are very difficult to enforce and rely on powerful forces being exposed. Once the opportunists gain momentum, they are very hard to derail.

This is why anticipation and early identification of shared values and like-minded, respected allies are so important in implementing values. If the pragmatist can make such an identification of the questionable practice and then recruit allies with similar values, the opportunist may be influenced to change course.

Although complicated, a deeper dive into what happened at Enron is necessary to fully understand how a values-driven lawyer might have helped avoid the catastrophe. Three links in the structure of critical financing transactions are important: (i) true sale opinions, (ii) total return swaps and (iii) sign off on quarterly and annual regulatory financial disclosures.

A chain is only as strong as its weakest link.

True sale opinions

In the case of the Enron "off balance sheet" transactions, the company used an accounting rule that required that a legal opinion (a "true sale" opinion) certify to the accountants that the value of an asset could be verified in a complicated financing by determining that the buyer assumed the risk of ownership and therefore would not pay more than fair value for the asset. The accountants would, upon receipt of that required opinion, issue financial statements that certified that the sale produced income and the financing would not be characterized as debt. The finance executives convinced the lawyers that they would avoid the "true sale opinion" and instead issue a "true issuance opinion." Although the "true issuance opinion" deceptively sounded okay, it was not the required document; it expressly acknowledged that it only covered the insubstantial transaction of the issuance of shares of the special purpose financing entity. Although the accountants bore responsibility for accepting this inappropriate document, the outside attorneys recognized that their opinions were facilitating the mischaracterization of the financing. After participating in this charade for a time, the lawyers took their concerns to their supervisors but were rebuffed.

Total return swap

In certain hedging transactions in commodity and energy transactions, the term "total return swap" is used to balance out the final settlement of the transactions. Those transactions do not purport to be sales of fixed asset values, but in Enron the term was used to disguise the guarantee of the repurchase of the asset that was the effect of the "total return swap" by the Enron parent company (which was, in itself, evidence that the guarantee repurchase of the allegedly sold assets should not be characterized as a sale for financial reporting purposes). The lawyers documented the transaction, including the "total return swap," knowing that this term disguised the guarantee. That deceit allowed the reporting of the transaction as income and not debt on the consolidated financial statements.

Quarterly and annual financial statements

The attorneys for the company approved the SEC's required annual and quarterly financial statements which exaggerated and mistimed the company's earnings and understated the company's debt obligations.

In each of these three circumstances, over multiple transactions and financial reports, company executives whose compensation bonuses and stock options were tied to earnings targets, pushed the lawyers to issue the deceptive legal opinions, disguise the effects of the misnamed total return swap and obfuscate the true nature of the financing transactions in the quarterly and annual regulatory filings.

The result was that the exposure of these gargantuan debts and the restating of the value of the assets and earnings caused the company to fail.

Lessons

In a lawyer's career and practice, conflicts of interest between broader constituencies (the public, consumers, defrauded or deceived parties) and client service are seldom as dramatic as the fiasco that was Enron. However, the values of honesty and integrity can often call out the opportunists before the issues spiral out of control as they did there. The ability to build allies, speak out when you see malfeasance and seek to place values as part of the objectives of representation may stop the abuse of opportunists who exploit situations for their economic or parochial interests in conflict with the people and broader interests they are employed to serve.

Could the Enron lawyer break the cycle of distortion and abuse employed by Enron financial executives to commit the fraudulent accounting that the special-purpose entity "financings" enabled?

In order to answer the question, the Enron environment should be dissected to provide context. First, the compensation system was based on a score card which evaluated performance based upon "creative" ideas that added to earnings. This incentive system rewarded all executives and attorneys who abetted the schemes that distorted the financial accounting. One creative idea was a plus, and two or three creative ideas were additional "good marks" toward higher cash compensation, bonuses and stock options. In the groupthink of "How can we channel new and novel approaches to the special-purpose entity financing?" cautionary comments that might slow the train were not encouraged. Imagine a large meeting where someone proposes a novel addition to the special-purpose entity formula that has an additional tax benefit or provides more front-end value, although with greater risk to the parent company guarantee, and someone wants to discuss the details of that risk. The naysayer in that context is likely to be dismissed and unwelcome at the next strategy meeting.

One important role of a lawyer as a professional is to identify risk. In this situation, there was risk to both the company (the real client) and to the executives and lawyers, as the subsequent collapse and bankruptcy demonstrated. How, then, could the lawyer find a forum to express both the value of honesty and the potential civil and criminal risks to which the finance executives were exposing themselves and the company in order to enhance their individual compensation ranks? Speaking up in the group was likely not a useful way to build alliances. But if the lawyer could selectively, individually counsel (i) executives on the finance side, (ii) the operating business unit that was subjecting its assets to the special-purpose entity financing and (iii) the general counsel's office attorneys who were charged with regulatory financial disclosure as well as corporate governance and ethics, perhaps the lawyer could begin to create a balance between the motive of increasing income as the sole objective and an appreciation of the individual and corporate risk being assumed as well as the corrosion of ethics and values being compounded by the multiple and increasingly risky special-purpose entity financing and refinancing.

Timing is important. Recognition of the imbalance between the goal of ever-increasing and consistently reported income with risk and value probably has to occur before the momentum of the fraud becomes self-sustaining and rationalized. In fact, one Enron lawyer, Jordan Mintz, actually sought a legal opinion from an uninvolved, major Wall Street law firm, on the efficacy of some of the financing techniques in an attempt to persuade the company of the risks. His efforts proved unsuccessful, but he tried, although likely too late in the process. Perhaps intercepting the phenomenon at an earlier stage, which would have required courage and prescience, may in a different company at a different point in time in the evolution of the scheme, yield a more positive result.

There are several specific lessons to be taken from the facts of the Enron debacle.

Switching sides

In many cases, lawyers who served as outside counsel to lenders/financiers or the company took jobs at Enron and, utilizing their knowledge of their former clients' and firms' procedures and their relationships with legal and non-legal personnel from the lenders and their former firms, negotiated extensions and increases in funds for transactions they had previously worked on. The abuse of trust built in those prior relationships led to increased bonuses and compensation for the individuals involved at the cost of millions of dollars of losses to the banks and their shareholders when the improper transactions defaulted. The trading in the trust relationships with their former colleagues was a corruption of such individuals' integrity and duties to their former clients and colleagues for personal gain.

A lawyer may give legal advice to a client based upon their subject matter expertise and experience in similar transactions, but that is very different from the lawyer switching sides on the same transaction (e.g. from lender's counsel to borrower's counsel). The Enron finance lawyers were skilled in structured finance, but that skill was not the sole reason that Enron bought their services when they were enticed to switch from lender's counsel or outside counsel to company counsel and worked on the same transactions. As described in the next chapter, the personal and professional relationship with their former client and colleagues created a trust relationship that reduced the diligence and scrutiny for amending and increasing funding for the same transaction that they had all worked on in the lawyer's prior role as the financier's counsel or outside counsel. Enron enticed the bankers in a similar fashion, but the lawyer is generally presumed to have a higher, independent duty to protect the client from fraud or mischaracterizations by abuse of legal opinions or structures. It was the abuse of this trust relationship for profit that corrupted the values of all involved.

Misuse of legal opinions

The Enron finance officials and in-house lawyers misused an accounting rule that was based upon the idea that a "true sale" of an asset was evidence of its value, and the income from such sale could be accurately reported as earnings. In order to qualify for such treatment, certain standards must be met. To prove that the standards were achieved, accounting principles required that outside attorneys, after due examination of the facts, issue a "true sale" opinion. The Enron business executives on the Finance teams instructed the outside lawyers to substitute a "true sale" opinion letter with a "true issuance" letter. This was worthless to accomplish the required test for reporting income but rather hid the underlying aspect of the transaction which was a guarantee of repayment (disguised as a "total return swap") which would, if recognized, cause the funds that were reported as income to be properly characterized as debt. The lawyers who were tasked to deliver the worthless "true issuance" opinions were afraid to speak up and honestly characterize the transactions due to the pressure to maintain the relationship with a huge client and encouragement from their former firm colleagues who had taken in-house legal positions at the company. Had the lawyers placed a higher standard on candor and honest expression of proper characterization of their roles as independent professionals, they may have exposed the opportunists.

Compensation

Enron had a policy that allowed the business executives to set the compensation of the in-house lawyers embedded in their units. Failure to comply with the dictates of the opportunistic Enron finance executives could cause the in-house lawyers to be terminated or receive reduced compensation, bonuses or stock options. There was no central legal authority to protect the overall values of the company; in fact, the only values at the finance units were to close the deceptive income generating transactions by any means necessary. The outside lawyers who attempted to report the misuse of the legal opinions and the deceptive characterization were silenced by their firms for much the same reasons.

Subsequent remedies

After the Enron collapse and its congressional investigations, litigation and bankruptcy investigations, corporate disclosure and whistleblower protection laws and regulations were adopted. No matter how large or complex a representation, a lawyer must develop and honor trusting, honest relationships with their clients by expressing from the initial meeting and the setting of the objectives of representation that the lawyer's values of honesty and fidelity to the client and its constituents, to fairness and candor, have equal weight with the profit motives of the client and the lawyer. The true purpose of the ethical rules, such as the legitimate purpose of client confidentiality, is to build trust, not to shield deceitful or fraudulent conduct.

Notes

1 Neal Batson, *Final Report of Neal Batson, Court Appointed Examiner* (New York: United States Bankruptcy Court, Southern District of New York, 2003), 26–52.
2 Neal Batson, *Final Report of Neal Batson, Court Appointed Examiner* (New York: United States Bankruptcy Court, Southern District of New York, 2003), 30–31.

Section 3

Core intangibles

Chapter 11

Trust and honesty

Trust

A client engages a lawyer to solve a problem or pursue an opportunity that usually represents a major life event, the outcome of which will greatly affect the client's life. Very often there is an important secret or crucial facts that are known to the client. The client is careful to guard the information because the revelation of these closely held emotions or facts make the client vulnerable to the counterparties to their transaction or dispute.

The lawyer brings expertise and diligence as well as the commitment to protect the client's closely held information. As described in Chapter 9, the ethics rules, legal rules and precedents create a framework for the lawyer to give the client assurance that they will maintain the client's confidences. However, without finding an alignment in the values of mutual loyalty and honesty, the parties may not form the essential bond of trust necessary for the lawyer to fully use their skills to meet the objectives of representation or have the ability to confidently express the lawyer's values.

Trust can be understood as the willingness to accept vulnerability because of positive expectations about the future behavior of another person.[1] Trust is the result of a firm belief in the reliability, truth, ability and strength of the lawyer and client in each other's commitment to the objectives of representation. If one party withholds crucial information regarding law, strategy or facts, the other party will be vulnerable to failures in accomplishing the best results.

Trust is often built over many shared experiences, which leads to reliability. But trust is more emotional than it is transactional.

Self-interest alone will not build trust. If one party has a deficit of self-interest, then the profitable exchange of self-interest is not enough to sustain trust because the party at the deficit of self-interest will feel vulnerable and question whether to continue to expose their vulnerabilities to others. However, when the cumulative experiences that bolster confidence in reliability are coupled with shared values of honesty, respect, fairness, responsibility and compassion, the vulnerable party will expect that the other party will treat their vulnerability in a caring way and build a reliable bridge to the next decision or choice to advance their agreed mutual objectives.

Betrayal is the opposite of trust. In a professional relationship of trust, both parties must intuitively feel that each can count on the other to not betray their vulnerabilities. The ethical rules and business practicalities alone cannot create the security of trusting the other to reliably and loyally protect each other and maintain the trust relationship.

The reliance on shared values as a complement to the ethical rules, legal protections and business calculations is the critical component to building and maintaining trust. The process of articulating and enacting shared values is the basis of the next important component of the lawyer being a trustworthy professional. That component is reputation. Over time, consistently reliable and trustworthy actions, as well as aspirations to enact personal values in all circumstances, will build a reputation among clients, colleagues, tribunals and adversaries that this person can be trusted. Trust in this way inspires and allows others to act in a trusting way and consistent with shared values.

Reciprocity is often associated with trust. If one party shows "good faith" and acts in a potentially vulnerable manner, then the other may reciprocate with another "good faith" act, signaling shared values of honesty and fairness. However, when one party seeks to take advantage of an offer of vulnerability and good faith, there is the temptation to retaliate/reciprocate with a similarly hostile act.

There is always the fear that trust will be abused, but reputations of others and experience in transactions and litigation will identify circumstances where trusting others is not warranted. Such warning signs may require greater caution but should not cause the lawyer to abandon their values in retaliation.

Abandoning your values has a corrosive effect on you and all that trust you. It also will not accomplish the objectives you seek to achieve. Lowering your values in retaliation for deceit or betrayal is like drinking poison and thinking your adversary will die.

In the long run, adherence to your values and serving as an example of aspirational conduct may not change the actions of your adversary, but it may inspire others with influence over the adversary (a judge, the adversary's clients or advisors) to exert influence to alter their actions. In most instances, when betrayal is recognized, steps can be taken to thwart or offset the harm of betrayal.

A guiding principle of *Giving Voice to Values* is that appealing to the fundamental values of honesty and fairness will attract allies and serve as a path to a resolution in a way that balances business, ethics and values to isolate the opportunist and allow the values orientation to find a place in the final resolution.

Honesty

Rule 4.1 of the Model Rules, *Truthfulness in Statements to Others*, requires that in the course of representing a client, a lawyer shall not (i) make a false statement of material fact to a third person, or (ii) fail to disclose a material fact when disclosure is necessary to avoid assisting in a criminal or fraudulent act by a client. Another rule (Rule 3.3) requires similar truthfulness ("candor") toward a tribunal or in an adjudicative proceeding.

My experience is that the perception of lawyers as liars is the most worrisome aspect of law practice for law students or young lawyers, who often express the fear that law practice will require them to "sell their soul" to be successful. Ask a practicing lawyer if they can go through their career without lying and there will likely be a long pause before a carefully constructed response, dissecting the context of the question and often contorting the definition of "lying."

Why is this so hard? Honesty is a core value and aspiration in a moral life. Lying is corrosive to a trust relationship and personal integrity. Assumption One of *Giving Voice to Values* is "the fundamental assumption that most of us want to find ways to voice and act on our values."[2]

To be "honest" is to be sincere and free of deceit. When the professional's actions do not reflect their true understanding of the facts and law or withhold important information necessary to faithfully respond to questions or situations which would impact the agreed objectives of representation, then the lawyer is not being honest nor fulfilling their ethical duties. Equally important, the lawyer in those circumstances is failing to uphold probably the singular most important value in the lawyer-client relationship: honesty. The Enron example shows the outcome of failure of honesty in a representation.

How does a lawyer improve the outcomes and reception that honesty may impact? First, they should have a plan of representation based upon their experience (or the gained experience of colleagues) which predict points in the representation when certain facts and legal objectives must be surfaced in the transaction or the litigation. Staging the revelation of troublesome facts or legal issues in a strategic manner is part of the skill of representation or negotiation. All facts and law need not be dumped out in a heap at the start of an interaction with counterparties, but a careful counseling relationship along with acknowledged shared values, especially including honesty, can help avoid surprises. Often the emotions engendered by surprise—caused by failure to plan and confirm values in early stages of representation—can cause abrupt responses that may not be truthful or at least distort facts or contain misleading information that needs to be corrected, thus undermining trust.

In preparing a plan of representation, the lawyer and client should agree on the staging of information in a manner that makes for the most productive, desired outcomes and builds the information into a context that will be most persuasive. For example, during the negotiation of price and terms of acquisition of real estate, the buyer may have certain characteristics that would appeal to a seller but also may inflate the seller's expectations as to the value the seller may extract from the buyer. Among the characteristics that the buyer may reveal (or conceal) are the buyer's ownership of adjacent property, the buyer's excellent credit or the buyer's control of a tenant for the property.

I once represented a client who sought to buy a property that had a very deep hole caused by prior surface mining. The mitigation cost of filling the hole by trucking in large quantities of dirt and fill over long distances using expensive heavy equipment greatly depressed the value of the strip-mined property. My client had very favorable relationships with trucking companies and had confidentially contracted to buy nearby properties with large mounds of dirt. Instead of revealing the buyer's special characteristics, we proposed a purchase contract with very substantial guarantees of financial performance in the form of an exceptional option payment up front. These guarantees allowed the transaction to proceed without disclosure of the buyer's other business arrangements. The buyer made a very successful transaction free of deceit or dishonesty by structuring information but buying the property at the seller's perceived value.

Trust built through honesty is fragile. One instance of deceit undermines trust, often irrevocably. However, the basis of honesty as a shared value is much stronger if it is based on a commitment to the value of honesty, not the fear of exposure or sanctions/penalties from others (including government and judicial reproach) or damage to reputation.

In other words, the aspiration of honesty (both for personal integrity and as an example for others) is an end in itself. It is not dependent on the truthfulness of opposing counsel, counterparties or others. It is possible that setting such an example may induce a higher level of trust and honesty in transactions, but whether that is the outcome or not, honesty should be an immutable part of the balance between ethics, business and values.

Honesty need not be brutal or aggressive. Tact and concern for feelings of others as well as skillful presentation of facts in their best light and in context with other factors are essential elements of persuasion. However, truthfulness itself cannot be finessed or rationalized. A compelling element of a values orientation is that we know when we are being honest and when we are not. Clients and colleagues often look for others to lie for them, under the assumption that lawyers serve outcomes, not values. Resist this invitation and follow the truth as you rightly know it.

As described in Chapter 10 regarding Enron, at certain points in a lawyer's career, they may face a client or adversary that elevates a monetary or social outcome ahead of telling the truth by purposefully misstating facts or legal precedence. The first presumption and projection of a values-driven professional is to suspect that the misstatement is an error or a small step "over the line" of truthfulness without malicious intent. However, it will often become obvious that the lie being advanced is mindful and told with malevolent deceit. Realizing what is happening, recognizing the perpetrator as a liar and calling out the lie, requires careful and steadfast support for the truth of the matter. Focus should be on the facts or the objective reality of the matter at issue. It is sometimes challenging to call out the lie, but if you marshal the facts and present the context and reality in as clean and supported a fashion as possible without getting pulled into name-calling, you will be able to create a persuasive rebuttal and not be dragged down to the level of the liar.

In the early stages of the Enron bankruptcy, there was an effort by the official Creditor's Committee, and the court appointed an Examiner to discover the facts behind many transactions that were improperly reported on the financial statements of the company. Meetings of Enron executives and lawyers for the Committee and the Examiner were convened with the responsible Enron business and finance executives and counsel for the court-appointed Restructuring Officer of the Enron Bankruptcy Estate ("Estate Counsel") seeking details and facts regarding the transactions.

The Enron executives were inclined to describe the transactions in a way that portrayed the transactions in positive, often glowingly optimistic, terms that did not adequately reveal all of the facts or potential financial harm to the counterparties to the transactions (most of whom were creditors of the Bankruptcy Estate). Additionally, the finance or business executives were accompanied in those meetings by criminal defense attorneys who represented the executives in their individual capacity (i.e. not as corporate executives).

When the descriptions of the facts by the Enron executives were seen to be misleading or deceptive to the Estate Counsel, the Estate Counsel would interrupt and describe the facts as he understood them based upon his knowledge from review of documents of the company and interviews with internal Enron personnel. The Estate Counsel's description usually portrayed the facts in a significantly less favorable light on the conduct of the Executive, the company or, in some cases, the counterparties by revealing omitted or mischaracterized facts.

When these descriptions occurred, there was an interruption of the meeting, and the executives and their criminal defense counsel called for a recess and a side conversation with the Estate Counsel. At that side meeting, the criminal defense attorney would angrily complain that the Estate Counsel should not reveal the facts that the Estate Counsel had obtained from the company records and interviews. The Estate Counsel then explained that his duty was not to the executive but to the creditors of the Bankruptcy Estate and also that the Estate Counsel's values required honesty and to seek the truth upon which fair remedies could be implemented in the bankruptcy proceeding.

Notes

1 Russell Korobkin, *Negotiation: Theory and Strategy*, 3rd ed. (New York: Wolters Kluwer Law & Business, 2014), 203.
2 Mary C. Gentile, *Giving Voice to Values: How to Speak Your Mind When You Know What's Right* (New Haven, CT: Yale University Press, 2010), 3. www.GivingVoiceToValues.org/

Section 4

Building a successful career

Chapter 12

Career choices

As a licensed professional, a lawyer may pursue many career paths and forms of employment, including (i) sole practitioner to association with other lawyers in a small, medium or large firm, (ii) government employment, (iii) judicial or regulatory roles, (iv) public interest, (v) academic pursuits and teaching and (vi) in-house work for business entities or non-profit organizations. In each instance, the duties to clients and the balance of business, client objectives (whether for profit or not), ethics and values are equally applicable regardless of the career choice or form of practice. It is likely that over the average 30–40-year career of a lawyer, they may pursue multiple choices of careers in the law or at least utilizing legal skills whether or not the lawyer's license is maintained.

Tom Robbins famously said "There are two kinds of people in the world, those who think there are two kinds of people and those who are smart enough to know better."[1] Similarly, the supposed dichotomy between business law and public interest law, or large firm versus small firm, or in-house versus large firm or any of the other choices does not materially affect the tension or stress that might be encountered when clients press the balance a professional should maintain between ethics, business and values. All forms of legal practice require the balance between these three elements. No career choice is a surefire insulator from the stresses of creating that balance.

Additionally, as described in Chapter 16, "The Role of the Lawyer in the Community," the form of legal career you choose does not preclude or exempt any lawyer from their obligations "to improve the law and the legal profession and to exemplify the legal profession's ideals of public service."[2] No matter the chosen career path, the protection of independence established by the government's issuance of a license to practice law carries with it the responsibility to be a "public citizen" and an example of the aspirational values described in this book.

The best place to start a career depends on several factors. First, where can the lawyer learn the most from skilled practitioners who are willing to exchange the young lawyer's diligence and commitment to learning with instruction in particular areas of the law that interest the young lawyer? However, as described in Chapter 14, "Role of a Mentor," a young lawyer should select their first job by discovering a senior lawyer with work habits, ethics and values that provide training and examples of balanced problem solving. The skills necessary to practice law are gained by the transfer of experience from other lawyers, exposure to challenging situations and problems of clients and interactions with courts, regulators and other lawyers. The counseling style, the work life balance and the legal organization's ideals and policies all present a varied landscape for a young lawyer to find a fertile environment to grow in their skills and experience.

The pace of change of the law and legal career opportunities are such that a lawyer typically needs to reassess their career choice every three to five years to decide if staying in their current position or making a change best suits growth, personal satisfaction, work-life balance and other financial and values-oriented goals.

In the past, lawyers chose a career path and stayed in that job for long periods of time. Mobility between jobs either within the environment of law firms or between law firms and government service or public interest (or any variation of these and other choices) is now the norm. Finding a specialized practice area that suits skills, values and personal life choices is much more common today and, in some instances, expected and encouraged. Transferring skills learned in one environment to another, especially in the continuing trend toward specialization rather than general practice, has become a typical career pattern.

Specialization can be an avenue for enacting values. There are many examples of personal passions and values guiding a lawyer to pursue a specialization. Many divorce or family lawyers have particular compassion for child development and creative solutions in family dissolution and child custody issues. Real estate attorneys may pursue a planning and zoning specialization in order to aid in shaping healthy and safe communities with parks and transit options that enhance sound environmental concerns and preservation of cultural and natural resources. Corporate practice specialties may focus on employment and retirement benefits that provide for a more positive future for older employees and retirees. A specialization in education law may allow an attorney to influence funding and sustainability of programs that allow students to find productive and creative paths to growth and future social contributions and affect educational accommodation for special needs students.

Each of these specializations can allow the attorney to serve the business interests of the client, meet the ethical standards of the profession and, importantly, implement their values to serve as an example of those values with the specialized client base that they serve.

Notes

1 Tom Robbins, *Still Life With Woodpecker* (New York: Random House, 1980).
2 American Bar Association Center for Professional Responsibility, *Model Rules of Professional Conduct* (Chicago, IL: American Bar Association, 2019), 2.

Chapter 13

Building a practice/ business development

Clients want three things from their lawyer: knowledge, experience and judgment. Satisfying those three requirements for a client is the foundation of building a law practice.

First, a client needs their lawyer to have knowledge of the applicable law and facts. All advice from the lawyer requires a firm grasp of the legal principles and options as applied to facts that have been or can be established.

Clients want a lawyer who is specialized in the legal problem they have retained the lawyer for. Specialization implies experience in similar transactions or litigation as well as specialized knowledge. A business owner selling a restaurant wants an attorney who has encountered the issues associated with such a sale and who has encountered the multiple issues (real estate, labor, trademark, financing, regulatory approvals) that accompany such a transaction. In a litigation setting, a lawyer's experience with the particulars of litigation (such as personal injury or breach of contract) can benefit a client by clarifying procedural aspects of discovery and pretrial motion practice as well as rules of court and temperament of judges or opposing lawyers. Very often the client will be encountering each particular legal process for the first time. In addition to the knowledge of the law and facts, the client wants the guidance of someone who has traveled this path in similar transactions or litigation before, someone who can help describe the time line and choices based upon prior experience.

Finally, and most critically, the client wants judgment. Judgment in this context is the skill of the lawyer to communicate the lawyer's knowledge and experience to accomplish the objectives of representation.

The bond between the lawyer and the client, and the satisfaction of both parties with the lawyer's performance, is largely dependent on the lawyer's ability to effectively communicate their knowledge, experience and judgment.

Business Development for lawyers is the ability to attract and retain clients; satisfying these three key requirements for clients is the foundation of that attraction and retention. As described in Chapter 3, the duty of diligence requires thorough command of the knowledge of facts and law. Experience is gained over time but can also be acquired by the creation of relationships with mentors and colleagues who have specialized in various legal and business specialties (e.g. labor law, real estate, trademark, personal injury litigation). Although personal experience is invaluable, the coordination of legal specialists by the attorney as the primary client contact is also an important aspect of client service, enhancing the ability to grow and expand legal representations and reputation.

However, the distinguishing factor between a representation based on just knowledge and experience is the addition of the lawyer's demonstration of their judgment. The likelihood of building enduring and satisfying legal relationships is correlated with the skill in merging all three factors of knowledge, experience and judgment. In order to sustain a lasting lawyer-client relationship, a client who relies on their lawyer as more than an expert or service provider must respect the lawyer's judgment and advice as to options, choices and direction of the representation.

In order to create a solid professional relationship as a lawyer that clients want to retain and rely upon—thus someone who can develop a growing reputation and practice—a lawyer must establish a bond that is dependent upon shared values.

Shared values are an essential element of the trust a lawyer and client must have in order to commit to the relationship; the client and lawyer should feel they are working together on a shared basis for a shared goal.

Introductions to clients or potential clients come through many avenues: introductions from satisfied clients, observation by opposing clients in litigation or transactions, social or community service encounters. In each instance, the client and/or opposing client is comparing the lawyer's knowledge and experience to that of the other counsel(s) involved in the litigation or transaction. Often the determining factor in switching counsel—and thus the key to building an expanded practice—is the judgment and values exhibited.

Economy and efficiency in thought and communication are hallmarks of the expression of judgment. There are a few maxims that should be kept in mind as effective ways to impress others with your judgment.

First, "listen to understand; don't listen to respond." A corollary to this principle is "think before you speak." Consider the objectives of representation in the context of the facts, particularly changing information or circumstances. A lawyer who gives good advice and displays good judgment is usually one who lets everyone involved express their views. By doing so, they come to understand the options presented and the expressed and unexpressed interests of the parties. After considering these factors, they propose a direction, often an adjustment, to take the litigation or transaction to the next phase with a prediction of the possible outcomes that are affected by the developments of information at hand.

Second, "experiences create judgment; bad experiences create good judgment." If possible, and within the bounds of client confidentiality, use past experiences in similar litigation or transactions (either personal experience or experience discovered from consultation with colleagues or research) to validate the advice you are proposing. Care should be taken to compare and contrast the prior relevant experience with the specific facts and law involved in your current situation, but it is reassuring and confidence-building to know that there is a path that has been traveled by others which yielded a desirable destination or outcome.

Third, judgment also includes forging a distinct new direction with full disclosure that although there is not a direct precedent, a creative solution may bring the parties to their desired outcome and the realization of their primary interests in a way that can create satisfying compromise. In other words, do not try to force a solution solely because there is a precedent, as the precedent alone may not consider all unique interests, facts or legal considerations of the current situation.

Fourth, take a position by expressing a preference for a recommended course of action. Clients want to understand their options and the basis for your recommended course of action, but they also want to know your advice on what they should do. Don't be dogmatic about your advice, but clearly take a position rather than just outline options.

Judgment is the synthesis of facts, law and precedents/experience; the ability to articulate solutions after careful listening; and effective, economical and efficient communication of a proposed solution. It is usually best for the solution to be expressed as a clear idea that can be repeated by all constituents in their own way of understanding the proposed action.

Finding common values in supporting the solution can often provide the foundation upon which business, economic and emotional factors can be expressed in a manner that all parties can subscribe to.

This book is about finding a higher meaning to law practice where the lawyer is more than a scrivener, a business problem solver or a tool in resolving controversy. Finding "success" in practicing law means creating outcomes for the client and lawyer that allow for the best and most satisfactory result to achieve shared values and objectives of representation after careful consideration of alternatives.

The concept of creating "successful outcomes" may be explained by the following example. A client approaches a lawyer who specializes in wills and estates. The client is married with two children and he asks the lawyer to draw up a "simple" will. The client also reveals that he had another child out of wedlock prior to marrying his wife. He has not told his family about the other child (to whom he secretly provides financial support) and asks for advice as to how to assume continued financial support for the "secret" child as part of his estate without revealing his existence to his family at this time.

There are mechanisms that could allow the client to protect his secret until his death, but the lawyer can ethically choose to offer advice on legal and values-oriented options that could provide a path for the client to examine his choice to maintain the secret. The lawyer has experience in similar situations as well as compassion for the client's choices and alternatives. The client's sharing of this vulnerability is an opportunity for the lawyer to be a confidant and build the support and trust that may enable the client to confront his choices and allow his family to come together in a healing, supportive way. The choice of revealing the "secret" child, and the effect on his family, can be analyzed together by the lawyer and the client in a safe, confidential space. The lawyer should, in my opinion, share the lawyer's values of honesty and openness but be mindful of the difficulties this revelation will have. The lawyer may recommend other professional counselors that could ease the process for the client, the family and the "secret" child.

In this case, the lawyer would encourage the client to bring the "secret child" into the open and provide a path to bring all family members into the recognition of the larger family structure in a manner that resolves the legal and psychological conflicts of the "secret" during the client's lifetime rather than having the surviving family members discover the deception at the death of the client. The lawyer in this instance chooses not to consider the representation as just an exercise in legal documentation. Here the lawyer accepts the invitation to go deeper by providing a comprehensive solution to the legal, financial and emotional issues presented. The lawyer takes the risk of vulnerability by accepting the consequences of negative outcomes of the client's choices but also creates the possibility of building a trusting, supportive relationship leading to a solution to a troubling and potentially very difficult outcome by continuing to maintain the secret.

Chapter 14

Role of a mentor

Values are internally generated by the individual and ethics are externally generated by the community the individual associates with. Judgment is often the intersection of values and ethics and, in the professional context of client representation and service, requires consideration of business needs and expectations. In this context, experience in various situations hones judgment and helps to anticipate conflict and create mutually beneficial solutions for all involved parties.

The practice of law is the transfer of experience.

Just as a client benefits from the perspective and distance of an independent, collaborative counselor, lawyers benefit from the experience and guidance of a mentor. The admonition to law students and lawyers entering the profession to "find a mentor" is a daunting task. The expectation that a mentor is a necessary part of professional growth may create pressure and unrealistic expectations on a young lawyer. The "assigned mentor" programs at law schools and legal organizations often fail to take into account an essential element and valuable aspect of mentor-mentee success: aligned values. A shared value system is a necessary component of the transfer of experience that can benefit a young lawyer.

A young lawyer need not have a single mentor but can access experience from different people in different contexts. A specialist in their chosen specialty may provide training and guidance of the law, procedure or custom in that specialty. An observant senior colleague may provide guidance on career choices, transitions, issues of balance or work habits. In order for this advice to resonate, the young lawyer must respect the values of the mentor and trust that the mentor is sharing their experience and wisdom in a generous way consistent with the foundation of shared values and transparency as to the mentor and mentee's motivations for seeking and giving advice.

There is an important difference between economic and social connections created in the typical student-teacher or employer-employee relationship and that of a mentor/mentee. Just because a senior lawyer is a skilled and successful practitioner or a teacher is a recognized expert or accomplished professional does not mean that they will fill a valuable advisory role as a mentor.

The mentor-mentee relationship requires more than just advice or predictions about outcomes in a case, transaction or career choice. The mentee should feel comfortable expressing their values and aspirations and the mentor should seek to understand those values and aspirations when providing advice that transfers the mentor's experience in a manner tailored to advance the mentee's aspirations consistent with their shared values in a generous manner (i.e. not in a manner that necessarily advances the interests of the mentor).

I would illustrate the previous point as follows: There have been many occasions in my career where an associate would come to me for advice or guidance about a career move or a legal or inter-office problem, the outcome of which would have some effect on my workload or my administrative duties within the firm. I always felt that it was incumbent on me to provide advice that I felt was in the best interest of the associate, as I had more power to deal with any adverse consequences to me. I wanted to treat the associate as I would have wanted to be treated in their situation by giving advice that identified their best interest and predicted outcomes or alternatives. My thought when an associate told me of a career or personal move that I knew was consistent with their values and aspirations (e.g. moving to a lower-stress atmosphere) was often "Good for you, bad for me." I always tried to explain that they should receive my advice as free of any self-interest on my part and not to conceal any part of their consideration that they worried would adversely affect our relationship. My view is that unless there is transparency, honesty and vulnerability (in essence "trust"), there won't be the full benefit of a mentor-mentee relationship.

It would be ideal if there were a completely detached mentor available for advice, but since the best advice requires familiarity with the context of the issues, usually a mentor has some continuing and preexisting relationship before critical advice is sought. The leap of faith by the mentee to seek advice on sensitive subjects requires trust that is often built through shared values and a sense of connection that is intuitive as well as created from past observations and interactions.

The mentor-mentee relationship may not rely solely on the disinterested generosity of the mentor. Once a young lawyer identifies someone whose values, skills and experience they respect, the approach for advice and ongoing connection as a mentor can be challenging. As in any successful relationship, there must be honesty, transparency and vulnerability. Once you express your desire for advice (vulnerability) and believe the other person will accept this request without using it to assert their strength, then the relationship can proceed in a positive manner.

There is another mentor-mentee paradigm that can also be very important. There is an inherent conflict in a mentor-mentee relationship in the same employment relationship. If the mentor is also the "supervising lawyer" for the mentee, then such mentor's responsibility to the firm may prevent appearance of favoritism or fair assessment of internal issues or divisions that could affect the advice given to the mentee or the candor the mentee can have with the mentor. Therefore, finding a mentor outside the firm where the mentee is employed but who has experience or skill in the community and/or practice area can also be a valuable resource. These "adjacent mentors" can come from various places, including bar associations, committees, multiple parties to transactions of litigations, family or community organizations. Perhaps in the interviewing process, the attorney encountered someone with whom they felt a connection but chose another firm. Maintaining contact and developing trust or confidence in the advice of these contacts may yield a valuable mentor.

Acknowledgment of vulnerability on both sides of the mentor-mentee relationship is, in and of itself, an expression of the values of honesty and transparency. The mentee acknowledges their lack of experience and need for guidance rather than "faking it." The mentor allows the mantle of authority which their status or experience brings to the relationship to be examined in discussions with the mentee.

It is also possible that a more traditional work relationship between an experienced lawyer and a less experienced lawyer can develop, through shared work assignments, into greater trust. That trust expands the transfer of experience from strictly a goal- client-centered work outcome to a more comprehensive exchange regarding career and other professional and personal challenges. Also, like all relationships, several attempts at finding compatible relationships may not be fully successful but can yield valuable insight into the qualities necessary to build trust and honest communication. In all events, professional life without the support and advice of more experienced colleagues and mentors is more difficult than being part of a supportive community and having a reliable mentor or mentors.

One method of identifying a mentor and building a mentor-mentee relationship may be as follows. An associate in a law firm identifies a partner in the firm that the associate believes has qualities that they respect and would like to emulate. The partner works in a different practice group from the associate. The firm encourages partners and associates to provide at least 50 hours per year for pro bono clients. The associate could ascertain the pro bono clients that the potential mentor partner is working for. The associate, through the pro bono coordination of the firm, finds a new opportunity presented by the partner's pro bono client and approaches the partner to request assignment to work with that partner on the new project. By working together, the associate will get to see the work habits and values orientation of the partner and hopefully build a relationship that could lead to mentoring opportunities and career advice.

Another mentor opportunity could be found in activities of the bar organizations or academic circles of certain practice specialties. If an associate interested in entertainment law were to attend the monthly luncheon of the local Bar Association Entertainment Section, they would meet more experienced lawyers in their desired specialization. Inquiries regarding potentially coauthoring articles on topics of mutual interest to the younger attorney and the more experienced attorney would present opportunities to build bonds that could lead to a mentor-mentee relationship.

In general, relationships built around working together on actual projects will lead to deeper and more lasting bonds upon which to receive the help of a mentor.

Chapter 15

Contributing to a cooperative workplace/ diversity

Values are exercised on both an individual and group level. Sophisticated representation on complex legal and business problems often requires the coordination of different legal skills. The alignment of values makes the application of varied skills more valuable to the client and more rewarding to the lawyers serving as colleagues on a team of lawyers, clients and other professionals.

A major impediment to the smooth functioning of a team seeking to solve legal and business problems is the insecurity visited on team members which is often caused by actual or perceived exclusion based upon factors that are not relevant to the skills that team member may have. Inclusion is critical to a well-functioning team. Inclusion and respect are the values that elevate a team member's contribution. If you feel insecure or not respected for your abilities, you will not feel a part of the effort of the team. Additionally, the anxiety and emotions associated with exclusion (i.e. not being fully embraced in your workplace or on a specific team) will impact your ability to be productive and feelings of job satisfaction and success will be impaired.

Inclusion is equated with the ideal of equality. Promotion of equality is a fundamental ideal of liberal values and a measure of progress. Respecting each individual in a diverse workplace facilitates inclusion and mitigates insecurity. Inclusion operates both institutionally and personally.

In any business environment, people bring factors other than their fundamental job skills. Such factors include economic status, personal goals and educational opportunities. Generally, legal environments bring bright, ambitious people together, but that similarity also brings differences in age, race, sexual orientation and religious beliefs that contribute to a healthy work environment by offering varied perspectives to empathetic problem solving.

Each person's position (lawyer, receptionist, paralegal, contractor) should be respected in order to allow them to be secure in the expectation that they can perform their best at their job and be evaluated on their job performance and delivery on the goals of the clients and institution based upon their contribution in their role on the team. Additionally, a cooperative, inclusive workplace provides each individual an opportunity for collaboration and a chance to learn, grow and improve their contributions and productivity in their job.

Therefore, each individual should commit to building a constructive workplace based on three principles: (i) respect, (ii) encouragement of growth and contribution and (iii) fair evaluation.

First, recognize and respect each person as a "valued team member" and respect their position as a partner, associate, paralegal, administrative assistant, receptionist or office services provider. Each person in their position deserves respect for the job they do and their contribution to the team.

Second, find ways to create open-ended opportunities for growth, continued learning and contributing to the institution's and clients' success and productivity. Organisms are either growing or dying. People who like repetitive tasks are rarely happy in a dynamic, successful legal environment.

Third, the institution and the individuals guiding it must provide fair and constructive evaluations that constructively criticize and compliment in an equal manner. Clearly explain compliments and criticism with specificity, both of past and desired future actions, with a positive outlook but a realistic assessment of obstacles.

The real test of a constructive workplace is the manner in which differences are acknowledged. First, recognize that every person is making a contribution within their job description. The values of inclusion and respect transcend the important legal restrictions such as physical accommodation for disability and prohibited discrimination based upon factors that are fundamental for each individual (race, sexual orientation, religion, etc.). Factors that will encourage inclusion and respect and will enable a constructive work environment on an individual level require balancing getting the work done (client needs) with respect for personal needs (family, health, religious observance, etc.). Another factor involves reciprocity. If one team member seeks an accommodation for family, religious observance or personal needs, the person making that accommodation should receive reciprocal accommodation for their needs as they arise. For entitlements to accommodation, a contribution to a constructive work environment should be sensitive to another team member's needs and offer reciprocal accommodation. The preceding is a long-term commitment and need not be case-by-case but rather an attitude of observation and empathy. Personal actions matter, and a good example set by a cooperative team member sets the tone for a culture of teamwork and building supportive relationships.

Collaboration and cooperation can be built and supported by assessing three questions:

1   How do we serve the needs of the client and complete the assignment?
2   What are the special needs and desires of the people affected?
3   What is the balance that is fair and acceptable?

The actions necessary to answer those questions should be:

1   Resolve to work with everyone; reduce awkwardness by focusing on work product.
2   In evaluations, focus on skills that serve the team and the client objectives without regard to other characteristics that are distinct to the individual.
3   Look for leadership qualities in everyone and expect the best. Don't be afraid to listen for good ideas and alternative solutions or to compromise.

Winning may be making a concession now without a clear benefit to you today. For example, an associate in a practice section does not work during several Jewish holidays throughout the year. The practice area expects several year-end closings in late December and has an annual rotation of allocation of vacation time over the Christmas holiday season. Since the observant Jewish associate's work is covered by her colleagues during the Jewish holidays, the

Jewish associate may waive her vacation allocation rotation during the Christmas holidays as an acknowledgment to her colleagues. This process does not require strict accounting of days worked versus absence but rather respects individual choices in a manner that shows a team concept to serving client needs and individual choices.

Chapter 16

The role of a lawyer in the community

Rule 6.1 of the Model Rules states: "[E]very lawyer has a professional responsibility to provide legal services to those unable to pay." This requirement of "pro bono" service is the most commonly described feature of the lawyer's ethical obligation to be a "public citizen." However, the role of a lawyer as a public citizen goes much deeper and is at the heart of a values orientation in the legal professional.

The Preamble to the Model Rules describes the depth of this responsibility. "A lawyer's conduct should conform to the requirements of the law, both in professional service to clients and in the lawyer's business and personal affairs."[1] Also, because the legal profession is largely self-governing through the ethical rules, the Model Rules prescribe that lawyers have a duty to "help the bar regulate itself in the public interest."[2]

Giving Voice to Values introduced the idea that values choices are constantly present. This constancy equally applies to the concept of the lawyer as a public citizen. As a member of the legal profession, a lawyer—in both their professional and personal life—should exercise their values in all their activities.

Instead of viewing these actions as duties or responsibilities, the more appropriate frame should be the values of generosity and gratitude. President John F. Kennedy paraphrased the biblical directive from Luke 12:48 as follows: "To whom much is given, from him much is expected." Gratitude for the opportunity to have the skills and resources necessary to serve clients, fellow citizens and other professionals should be a powerful motivation for values-oriented actions in all aspects of a lawyer's life, whether as a parent, a school board member, a teacher or a practicing attorney. Generosity of spirit, giving without the expectation of return is a gift of such public service in its broadest sense. A gift is as valuable to the giver as the recipient. The example set by a lawyer who embraces the values of generosity and gratitude in their professional and private life will enhance their life and set a powerful example for those affected by such gratitude and generosity.

Prior chapters have examined the directional relationship in a professional's career in the context of the transfer of experience from a mentor to a mentee and the cooperative relationship between team members in a workplace. Both of these essential foundations of a successful, growing and gratifying law practice are sustainable not through raw self-interest but rather by the infusion of gratitude and generosity. Self-interest or duties and responsibilities of ethical codes or social custom will not provide the same satisfaction or reliability as implementation of the genuine values of gratitude and generosity. Practicing these values will not only improve with habitual exercise and create personal and professional satisfaction but will also inspire others to act in the same way, first through the social heuristic of reciprocity but eventually through the satisfaction of living the aspiration of a values-oriented frame of action.

Notes

1 American Bar Association Center for Professional Responsibility, *Model Rules of Professional Conduct* (Chicago, IL: American Bar Association, 2019), 1.
2 American Bar Association Center for Professional Responsibility, *Model Rules of Professional Conduct* (Chicago, IL: American Bar Association, 2019), 2.

Index

Printed in Great Britain
by Amazon